W0106673

POLYMER SCIENCE AND TECHNOLOGY
Volume 18

# REACTION INJECTION MOLDING AND FAST POLYMERIZATION REACTIONS

# POLYMER SCIENCE AND TECHNOLOGY

## *Recent volumes in the series:*

A Continuation Order Plan is available for this series. A continuation order will bring delivery of each new volume immediately upon publication. Volumes are billed only upon actual shipment. For further information please contact the publisher.

POLYMER SCIENCE AND TECHNOLOGY
Volume 18

# REACTION INJECTION MOLDING AND FAST POLYMERIZATION REACTIONS

Edited by

## Jiri E. Kresta

*Polymer Institute*
*University of Detroit*
*Detroit, Michigan*

PLENUM PRESS • NEW YORK AND LONDON

Library of Congress Cataloging in Publication Data

International Symposium on Reaction Injection Molding (1981: Atlanta, Ga.)
  Reaction injection molding and fast polymerization reactions.
  (Polymer science and technology; v. 18)
  "Proceedings of the International Symposium on Reaction Injection Molding, spon-
sored by the American Chemical Society Division of Organic Coatings and Plastics
Chemistry, held March 31 – April 1, 1981, in Atlanta, Georgia"—Verso t.p.
  Bibliography: p.
  Includes index.
  1. Plastics—Molding—Congresses. 2. Polymers and polymerization—Congresses. I.
Kresta, Jiri E.. Date –         II. American Chemical Society. Division of Organic
Coatings and Plastics Chemistry. III. Title. IV. Series.
TP1150.I57 1981                    668.4′12                      82-12389

ISBN 978-1-4684-8735-0   ISBN 978-1-4684-8733-6 (eBook)
DOI 10.1007/978-1-4684-8733-6

Proceedings of the International Symposium on Reaction Injection Molding, sponsored
by the American Chemical Society Division of Organic Coatings and Plastics
Chemistry, held March 31–April 1, 1981, in Atlanta, Georgia

© 1982 Plenum Press, New York
A Division of Plenum Publishing Corporation
233 Spring Street, New York, N.Y. 10013

Softcover reprint of the hardcover 1st edition 1982

PREFACE

The emergence of reaction injection molding (RIM) has been followed by the industry with mounting interest. RIM technology has brought to polymer processing a new flexibility and savings in both energy and capital investment. The new developments and the number of engineers and scientists working in RIM is growing at so fast a rate that there is need for sharing information on progress in this area.

This book is based on papers presented at the International Symposium on Reaction Injection Molding which was held in 1981 in Atlanta, Georgia, and was sponsored by the American Chemical Society, Division of Organic Coatings and Plastics Chemistry. The book is divided into four parts covering different areas of RIM development.

The first part is devoted to the future trends of RIM development in the United States and Japan.

The structure-properties relationship and effects of annealing on properties of RIM elastomers are covered in the second part.

New non-urethane polymers such as polyamides, polyisocyanurates and polystyrene suitable for RIM processing are discussed in the third part.

In the last part the engineering and technological aspects of RIM, such as glass reinforcement, mixing, flow and moldability are covered in detail.

Finally I would like to thank Mrs. Iris Glebe for typing this book and for help with editing, and K. Zielinski for his assistance

in reviewing and, when necessary, correcting some of the papers.
Thanks is also due to the editors of Plenum for their patience and
helpfulness.

Jiri E. Kresta

Polymer Institute
University of Detroit
Detroit, Michigan 48221

CONTENTS

FUTURE OF RIM DEVELOPMENT IN THE U.S.A. IN THE 1980's

Louis M. Alberino,

The Upjohn Company
D. S. Gilmore Research Labs
North Haven, CT 06473

INTRODUCTION

This paper covers the development of RIM in the U.S.A. to the present, as well as the future of RIM and RRIM in the U.S.A. in the 1980's.

RIM molded urethane elastomers for automotive bumper (fascia) covers were first commercially produced in the 1975 model years on the General Motors' Monza, Skylark and Starfire series. This introduction of commercial RIM molded articles has been followed by a steady growth in use, particularly with elastomeric RIM, resulting in the use of some 50 million lbs. of RIM elastomer, most of which went into the automotive market in 1980 (Table I).[1] Rigid RIM use amounted to approximately 20 million pounds (Table I) and along with non-automotive elastomeric RIM represents a new area of growth for RIM.

This paper discusses some of the details of the above markets (Section II) since they act as the stimuli for the present and future developments which are discussed in Sections III, IV and V.

RIM AND RRIM MARKETS IN THE U.S.A.

The projected RIM usage for 1985 shown in Table I can be further examined not only by the type of RIM material (rigid vs. elastomeric) but also by the market and use, such as automotive market. Table II shows that although the average weight of American cars will be further downsized from 4200 lbs. in 1979 to 3000 lbs. in 1985, total plastic usage will increase (since plastic usage plays a key role in

1

Table I.  Projected RIM Usage by Type[1]

|  | Millions of Pounds | |
|---|---|---|
|  | 1980 | 1985 |
| Rigid RIM | 15 | 40 |
| Elastomeric RIM | 50 | 110 |

weight reduction). Total urethane usage[2] also is expected to increase by 1985. The breakdown for the various types of urethanes is given in Table III.[2] The designation foam in Table III includes slabstock, molded seating, bumpers and head restraints. Some of these may be RIM molded but others (e.g., slabstock) most likely are produced using low pressure equipment. Note the predicted large growth in RIM elastomer and the appearance of RRIM (scheduled for the 1981 Sporty Omega as RRIM molded front fenders.[3] Thus RRIM automotive body paneling parts represent a new application area for RIM. The automotive elastomeric parts consist of bumper covers, full front fascia steering wheels, sight shields, air dams, deflectors and spoilers of which bumper covers and fascia account for approximately 90% of the poundage (as of 1980).

Some of the elastomeric RIM and most of the rigid RIM reported in Table I goes into non-automotive applications. Some of these applications are described in Table IV. Many of these applications are expected to contribute significantly to the growth of RIM in 1985 and the latter part of the 1980's.

THEORETICAL UNDERSTANDING

As in so many other areas, large scale commercial use of RIM preceded a full understanding of the technology from a theoretical point of view. However, as the surge into new applications contin-

Table II.  Plastics Usage vs. Car Weight[2]

| Model Year | 1979 | 1985 |
|---|---|---|
| Average Weight, lbs. | 4,200 | 3,000 |
| Total Plastics, % | 4.86 | 7.66 |
| Total Urethane, % | 1.03 | 1.83 |

Table III.  Present and Future Use of RIM[2]

Urethane Elastomers and RRIM in American Automobiles

| | Millions/Pounds | |
| Model Year | 1979 | 1985 |
| --- | --- | --- |
| RIM Elastomer | 37.4 | 116.8 |
| RRIM | - | 69.8 |
| Foam* | 269.2 | 250.4 |

*Includes production via low pressure processes as
well as RIM.

Table IV.  Non-automotive RIM Applications

| Application | RIM Type |
| --- | --- |
| Business machine housings | rigid |
| Recreational (ski boots, golf carts) | elastomeric |
| Recreational (ski cores, water skis) | rigid |
| Shoes (soling) | elastomeric |
| Agricultural | elastomeric |
| Appliances | elastomeric/rigid |

ued, researchers at various academic and industrial labs began to examine the RIM process from a more fundamental point of view.  This
research can generally be divided into a study of the RIM process
itself and a study of the polymer produced.

The RIM process itself has been divided into three component
parts, impingement mixing,[4] mold filling,[5] and curing (kinetics).[6,7]
These have recently been reviewed.[8]  The results of these and other
investigations show that a Reynolds number of 200 will give good mixing and that the mold should be filled before large pressure buildup
occurs.  This has become more critical in recent years with the faster

reaction systems, some of which are discussed in Section IV.

The above studies were conducted primarily on elastomeric or non-rigid type RIM materials which may be only slightly blown (sp~1.0). Rigid RIM generally has been involved in more highly blown systems (sp~0.6) and in these instances the foaming process is also a critical part of the analysis.[9]

The understanding of the RIM process gained through the above studies has helped to contribute towards further improvements in RIM equipment, some of which are discussed in Section V.

Since RIM produced elastomeric polyurethanes are segmented two phase polymers, the basic studies which have been done of the morphology of TPU's and cast materials apply as well to RIM elastomers. A good review of these results can be found in Reference 8. In order to achieve complete reaction and yet useful properties, the right balance of compatibility and incompatibility between the chemicals and the resulting polymer must be achieved.[10,11] For rigid RIM one of the key factors is the balance between the foaming process and the chemical reaction.[9] Each of the types of RIM materials demands special attention to the balance between processing and properties.

Because of the variety of formulating possibilities, a range of products can be achieved. This range of materials possible is listed in Table V. As shown in Table V, the newer applications in the future will be coming from the high modulus materials which are not true elastomers.


CHEMICAL DEVELOPMENT

Currently, commerical RIM produced articles are based on isocyanates, polyols, extenders, catalysts and blowing agents. The extenders are either glycols or amines or some combination of the two. The major chemical producers of isocyanates and polyols for RIM use are listed in Table VI. The total urethane market in North and South America is expected to be 3.6 billion pounds in 1983 (of which elastomers account for 363 million pounds).[12] Isocyanate expansion has kept pace, and MDI (crude and pure) capacity in 1983 will be approximately 800 million pounds. The pure MDI types are used in elastomeric RIM and some higher modulus RIM, while the crude MDI types are used in rigid RIM. It is expected that MDI will still be the prime isocyanate for isocyanate based RIM in the near future, with the aliphatics being used for light stable materials.[13] Some efforts are being put into non-phosgene routes to MDI, although these have met with limited success to date.[14]

Because isocyanate technology lends itself to the production of a variety of MDI based materials, and since polyols themselves are

Table V

General Properties of Isocyanate Based RIM Systems

| | Low Flexural Modulus | Intermediate Flexural Modulus | High Flexural Modulus |
|---|---|---|---|
| Flexural modulus, kpsi @ RT | 20 – 75 | 75 – 150 | 200 – 400 |
| Tensile elongation, % @ break | 100 – 300 | 50 – 200 | < 50 |
| Izod impact, ft-lb/in | 10 – 15 | 5 – 15 | < 5 |
| Impact strength | high | medium – high | low |
| Material description | elastomer | pseudo-plastic | plastic |
| Application (automotive) | fascia | fender | hood or deck lid |

Table VI.  Major Producers of Isocyanates and Polyols

|  | Isocyanate (MDI) | Polyol |
|---|---|---|
| BASF Wyandotte Corporation | yes | yes |
| Dow Chemical Company | - | yes |
| ICI (Rubicon) | yes | - |
| Mobay Chemical Corporation | yes | yes |
| Olin Chemical Company | - | yes |
| Texaco Chemicals | - | yes |
| Union Carbide Corporation | - | yes |
| The Upjohn Company | yes | - |

available in a variety of types, polyurethane materials will be further developed in the future.  The next two sections discuss future developments in urethanes and other isocyanate based materials.  The final section concludes with a discussion of non-isocyanate based RIM.

Polyurethane Development

The present state of polyurethane development has been a result of efforts by isocyanate, polyol and extender suppliers.  Over the past several years various liquid versions of MDI have been developed for use in RIM.[15]  Polyol development has centered on the end-capping of polypropylene glycol with EO to yield high primary OH terminated polyols with good reactivity.[16]  Recent polyol developments have been the grafted types[17] and those with other "hard segment" blocks internal to their structure.[18,19]  Further development in polyols would appear to involve further efforts related to incorporating various types of "hard segments" into the polyol backbone.[20]

Although the use of aromatic amine extenders would give aromatic urea "hard segments," traditionally these extenders have been considered along with glycol as "urethane" extenders.[21]  Although their use in urethanes was well established in other areas (cast elastomers and foams,[22,23]), the use of amine extender in elastomeric RIM is relatively new.[24]  Said to impart fast demolding and good green strength,[24] it is expected that further developments with amine extenders will occur, as is evident from the volume of recent patents on the subject.[25,26,27,28,29,30]

The very reactive isocyanate function also can react with itself in a trimerization reaction to produce the isocyanurate structure.[31] This technology can be combined with urethanes to produce very high temperature resistant plastics.[32] Although these are brittle plastic materials because of their high crosslinking density,[33] they may find use as suitable binders for long-fiber reinforcement in a mat-molding type of process.[34]

## Non-urethane Isocyanate Based Polymers

Isocyanates can undergo other reactions besides the ones described above. Often used in combination with glycol extenders, or polyols, these reactions can produce new structures such as polyamides using the presently developed RIM process.[35,36] Additionally, as newer types of equipment are produced (see Section V), other reactions of isocyanates may be put to further development in RIM type processing.[37]

## Non-isocyanate Based RIM

Recently, a great deal of attention has been given to RIM processing non-isocyanate non-urethane systems. Chief among these have been polyester,[38] epoxy[39] and Nylon 6.[40]

The processing of caprolactam based systems probably represents one of the more promising approaches because of the economics and the physical properties of the resulting polymer. Potentially polyester is the least costly; however, the styrene content may pose a problem, although it may be overcome by processing polyester with very low styrene contents.

## MACHINERY DEVELOPMENTS

In general, the RIM process consists of the RIM metering equipment itself with its storage tanks, as well as clamps or presses to hold the tool in which the molding is to be made.[1]

There have been recent developments in all of these areas, as well as new designs in RIM equipment. Improvements to conventional RIM equipment have included new mix heads[41,42] both for unfilled and filler (RRIM) RIM. Following the lead of thermoplastic injection molding, microprocessors and other types of electronic instrumentation and controllers have been incorporated into the feature of the latest offerings in RIM equipment.[43,44]

The advent of filled RIM, generally termed RRIM, was brought about through advances in equipment.[45] It appears just as likely that further advances in RIM design will give rise to new generations of systems such as the equipment development now going on with Nylon

6.[40] However, it should be pointed out that this new generation of equipment (heat traced lines, heated pumps, perhaps higher pressures, etc.) can also be used with chemical systems heretofore not possible with conventional RIM. Thus, there will be increasing numbers of candidate systems with perhaps economics being the decisive factor in the choice.

CONCLUSIONS

In summary, the future of RIM in the U.S.A. will continue to be fueled by automotive demand and applications. However, an increasingly large market will develop in non-automotive applications.

The types of materials will include not only the elastomer type but also the higher modulus types. These developments will be backed by an increasing use of new polymer types made possible by new and radical developments in the RIM equipment itself.

## References

1.  "Reaction Injection Molding," W. E. Becker, Van Nostrand Reinhold Co., New York (1979).
2.  D. G. Leis, SPI/FSK Int. Conference, Strasbourg, France (1980).
3.  "Plastics Design Forum," 6 (1), January/February, 1981, p. 12.
4.  L. J. Lee, et al, 37th ANTEC, SPE Tech. Papers, 439 (1979).
5.  J. M. Castro, et al, 37th ANTEC, SPE Tech. Papers, 444 (1979).
6.  S. D. Lipshitz and C. W. Macosko, J. Appl. Polym. Sci., 21, 2029 (1977).
7.  E. C. Steinle, et al, J. Appl. Polym. Sci., 25, 2317 (1980).
8.  L. J. Lee, Rubber Chemistry and Technology, 153 (3), 542 (1980).
9.  "Reaction Injection Molding," Chapter 5, W. E. Becker, Van Nostrand Reinhold Co., New York (1979).
10. R. J. Lockwood and L. M. Alberino, "Organic Coatings and Plastics," Chem. Preprints, 43, 899 (1980).
11. R. B. Turner, et al, 181st ACS Meeting, Atlanta, GA (1981).
12. The Upjohn Co., Polymer Chemicals Division.
13. J. Verwilst and E. DuPrez, SPI/FSK Int. Conference, Strasbourg, France (1980).
14. U. S. Patent 3,919,278 (1975) ARCO.
15. The Upjohn Co., Polymer Chemicals Div., Technical Service Report TS-1.
16. H. J. Fabris, "Adv. Ureth. Sci. and Technol.," 3, 108 (1974).
17. Technical Bulletin, "Niax Polymer Polyols," Union Carbide, NY.
18. B. A. Phillips and R. P. Taylor, Rubber Div., ACS, Atlanta, GA (1979).
19. European Patent Application 79102268.4 (1979), Dow Chemical Co.
20. L. J. Lee, ibid, p. 568.

21. J. H. Saunders and K. C. Frisch, "Polyurethanes, Chemistry and Technology, Part I, Wiley, 1962, New York.
22. P. Wright and A. P. C. Cumming, "Solid Polyurethane Elastomers," Gordon and Breach, New York (1969).
23. U. S. Patent 4,048,105 (1977), McCord Corporation.
24. R. P. Taylor and B. A. Phillips, SAE Int. Congress Exposition, Detroit, MI, 1981.
25. U. S. Patent 4,202,987 (1980), Ciba-Geigy Corporation.
26. U. S. Patent 4,208,507 (1980), BASF Wyandotte.
27. U. S. Patent 4,181,682 (1980), Texaco.
28. U. S. Patent 4,229,561 (1980), Lim-Molding, S.A.
29. U. S. Patent 4,246,425 (1981), Ihara Chemical Industry.
30. U. S. Patent 4,174,240 (1979), Bayer.
31. W. J. Farrissey, L. M. Alberino and A. A. R. Sayigh, J. Elast. Plast., 7, 285 (1975).
32. P. S. Carleton, D. J. Breidenbach and L. M. Alberino, 1979 NATEC, SPE (1979), Detroit, MI.
33. L. M. Alberino, J. Appl. Polym. Sci., 23, 2719 (1979).
34. H. N. Marsh, et al, Soc. Plastics Ind., Tech. Conf., 34, 3-A (1979).
35. D. F. Regelman and L. M. Alberino, "Organic Coatings and Plastics Chemistry," Atlanta, GA (1981).
36. D. F. Regelman and L. M. Alberino, ibid, Atlanta, GA (1981).
37. W. J. Farrissey, L. M. Alberino and A. A. R. Sayigh, ibid.
38. Plastics Technology, p. 10, October, 1978.
39. R. S. Kubiak and R. C. Harper, SPE 35th Annual Technical Conference, 22-C (1980).
40. Plastics Technology, 27 (4), 29 (1981).
41. Modern Plastics, p. 38, May, 1980.
42. R. C. Harper and D. M. Reber, Soc. Plast. Ind., Tech. Conf. 673 (1979).
43. Modern Plastics, p. 52, December, 1980.
44. Modern Plastics, p. 50, February, 1981.
45. L. J. Lee, ibid, p. 545.

# RIM AND RRIM DEVELOPMENT IN JAPAN

Kaneyoshi Ashida

Mitsubishi Chemical Ind., Ltd.
5-2, Marunouchi 2-chome
Chiyoda-ku, Tokyo, Japan

## INTRODUCTION

The use of RIM products in the automotive industry in Japan was started by the stimulus of the successful use of RIM bumpers in the United States. The first use of RIM bumpers in Japan was seen in the Toyota Celica in 1977 and was followed by Mitsubishi cars in 1978, Nissan (Datsun) cars in 1979 and Mazda cars in 1980. The major use of RIM in the Japanese automotive industry, therefore, is in the production of bumpers. To a lesser extent it is used in such automotive parts as dash boards, steering wheels, etc. The major non-automotive uses of RIM are in ski cores and computer housings. The total usage of automotive RIM parts in 1979 amounted to approximately 7,000 metric tons; total usage of non-automotive RIM or rigid RIM foams amounted to approximately 1,000 metric tons. The consumption of various RIM products in Japan in 1979 is shown in Figure 1.

The rise in the annual production of motor vehicles in Japan for the last 2 decades since 1960 is shown in Figure 2. In 1980 the production of passenger cars reached 7 millions for the first time. The shortage of oil and its subsequent high cost resulted in severe regulations in improving the fuel efficiency of cars. Table I shows a joint notification of the Ministry of International Trade and Industry (MITI) and Ministry of Transportation regarding the fuel efficiency of cars. The notification was issued in December, 1979, as a guideline for improving fuel efficiency of cars. It requires that improved mileage be reached by 1985. For example, a car having an equivalent inertia weight of 1,000 kg must meet the mileage requirement of 12.5 km/$\ell$ or 29.3 mpg. This target is much higher than the target in the U.S.A., i.e., 11.7 km/$\ell$ or 27.5 mpg.

11

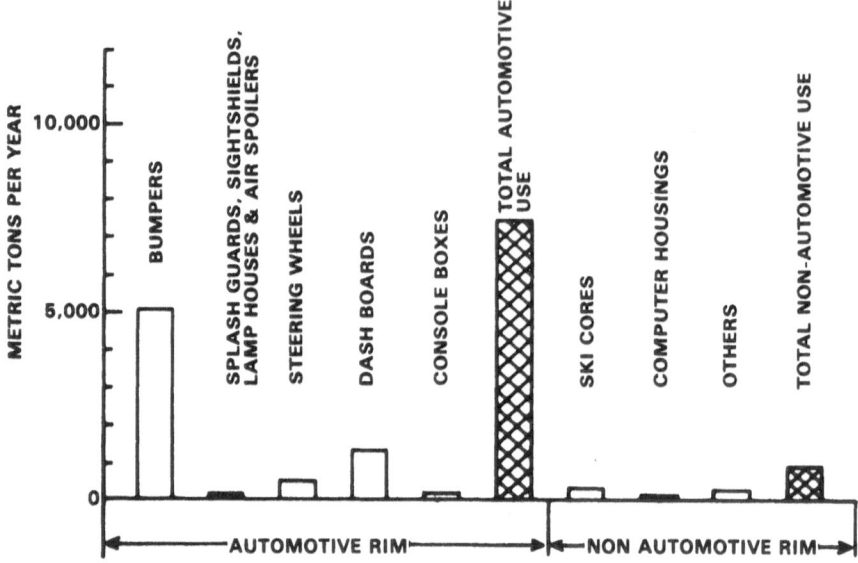

Figure 1.   Consumption of Various RIM products (1979).

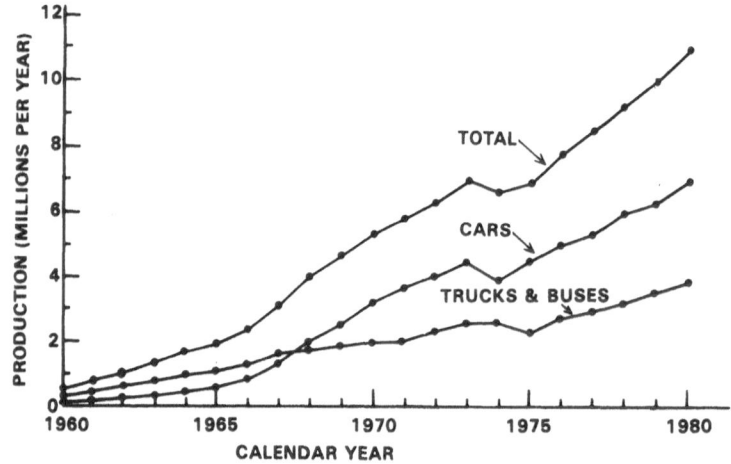

Figure 2.   Motor vehicle production in Japan (source:   JAMA)

In order to meet these requirements, the reduction of car weight is strongly suggested.

Since Japanese cars have already been down-sized, improved fuel efficiency could be achieved only by down-weighting of cars, and the down-weighting can be achieved by the increased use of plastics, high

Table I. Notification Concerning Fuel Efficiency. (December 27, 1979, MITI/Ministry of Transportation.

| EQUIVALENT INERTIA WEIGHT | | FUEL CONSUMPTION-1985 | |
|---|---|---|---|
| KG | (LB) | KM/L | (MPG) |
| 577.5 | 1271 | 19.8 | 46.5 |
| 577.5-827.5 | 1271-1821 | 16.0 | 37.5 |
| 827.5-1265.5 | 1821-2784 | 12.5 | 29.3 |
| 1265.5-2015.5 | 2784-4434 | 8.5 | 19.9 |

tensile steel and aluminum in place of regular steel. A similar trend can be seen in American and European cars. Some efforts in improving fuel efficiency, e.g., computer-controlled engines and other innovative engines are being developed, but these are outside the scope of this paper.

Another factor influencing the future design of cars lies in the severe emission standards in Japan. Japanese cities are densely populated, and traffic is very heavy; therefore, Japan has very severe regulations in emission standards.

An increasing use of plastics for Toyota cars in comparison with American and European cars is shown in Figure 3. The weight of plastics per car in Japan is almost equal to that of European cars, due to their similar sizes. In contrast, American cars employ much more plastics than Japanese and European cars. This is due in part to the larger size of American cars.

The use of plastics per car in 1979 and its estimated figures for 1980 are shown in Figures 4 and 5. PVC followed by PP and polyurethane are the plastics used most in the automotive industry. In fact, all plastics show an increase in use except ABS. This picture is different in the U.S.A. where PVC is the third most widely used of automotive plastics.

The estimated figures of the use of polyurethanes per vehicle in 1980 are shown in Figure 6. It is projected that about 9 kg per car, 5 kg per truck, 20 kg per bus and 0.75 kg of polyurethane per two-wheelers will be used.

The estimated figures for the total motor vehicle use of all types of polyurethanes in 1980 is shown in Figure 8. It is clear from these figures that the biggest market for polyurethane foams lies in transportation, and its consumption in 1980 is estimated to be over 50% of the total flexible polyurethane foam used. This paper discusses the current and future trends of RIM and RRIM usage in both automotive and non-automotive industries in Japan.

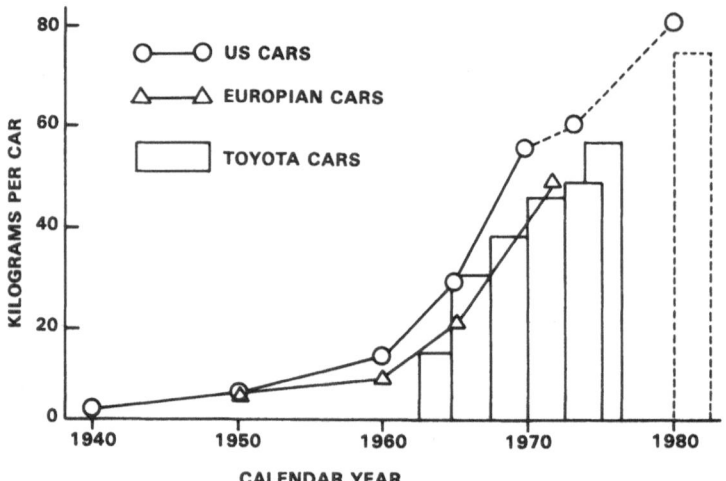

Figure 3.    Automotive plastics in Toyota cars
(Source:   Polymer Preprint, Japan)

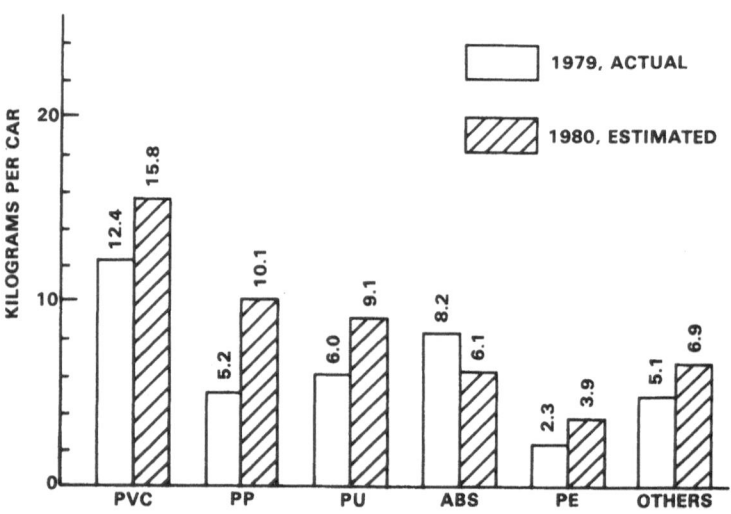

Figure 4.   Use of plastics per car (source:   JAMA)

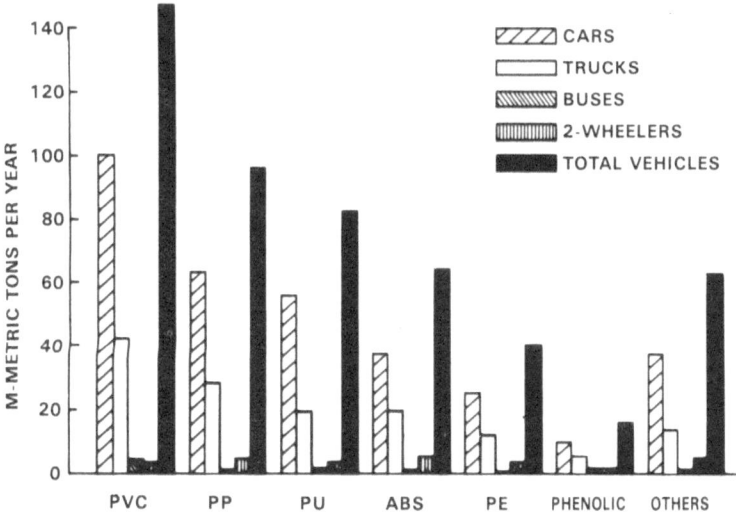

Figure 5.   Automotive plastics - 1980 (estimated).   (Source:   JAMA)

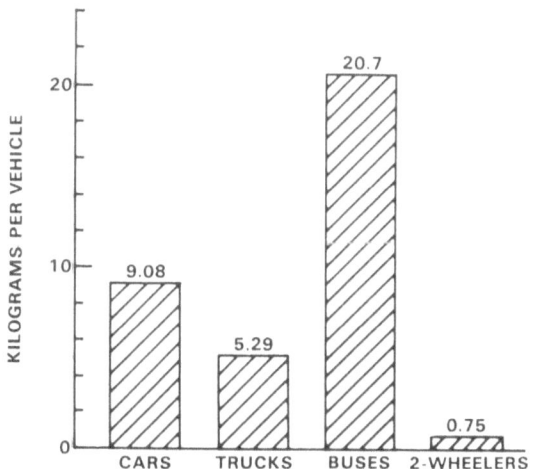

Figure 6.   Use of polyurethanes per vehicle -
            1980 (estimated).   (Source:   JAMA)

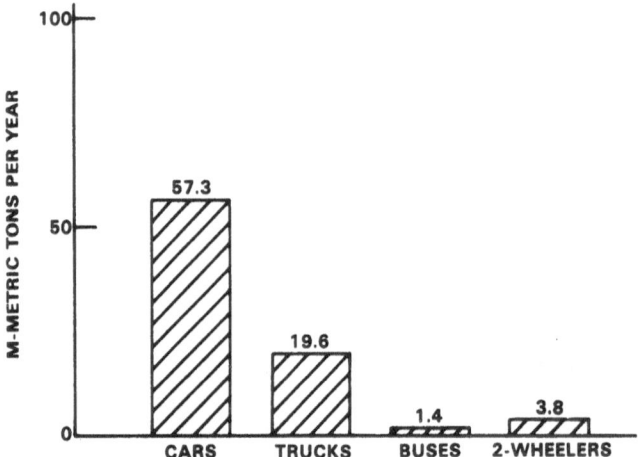

Figure 7.   Motor vehicle use of polyurethanes
            in 1980 (estimated).  (Source:  JAMA)

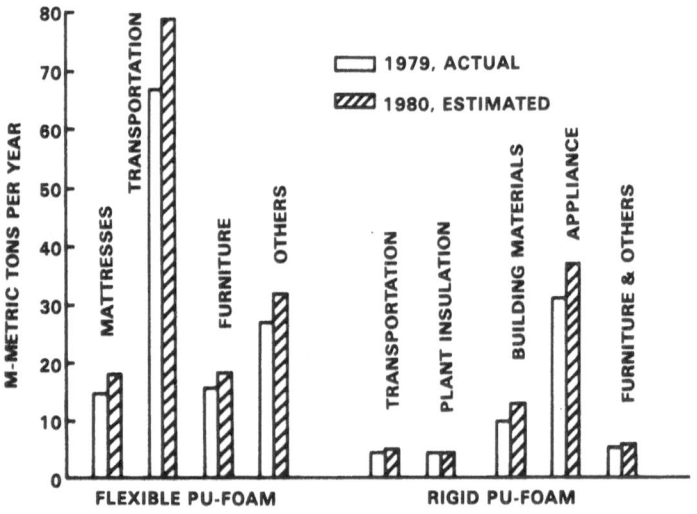

Figure 8.   Use of polyurethane foam (1979–1980)
            (Source:  Sekiyu Kagaku Kenkyusho)

AUTOMOTIVE USE OF RIM POLYURETHANES

## Bumpers

The types of bumper systems employed in Japanese cars are shown in Table II. They are: (1) RIM cover + RIM foam; (2) RIM Cover + steel beam + hydraulic system; and (3) RIM cover + plastic honeycomb. These types can meet the requirements of FMVSS 215 and Cost Saving Act, Part 581. Type (1) is the most popular and promising in Japan, and is preferred by Toyota and Mitsubishi.

An important point to identify is that energy-managing foam may absorb water and may be frozen at sub-zero temperatures, which results in the loss of shock-absorbing properties. In spite of this disadvantage, the system is extensively used in Japan, presumably because of the difference in weather conditions between the U.S.A. and Japan. In addition, a combination of the RIM cover and steel beam cannot meet the requirements of the FMVSS regulation, but this does not deter its use in Japan and western Europe.

RIM bumpers for the 1981 model Toyota cars are shown in Table III. All Toyota bumpers are composed of a RIM bumper cover and RIM energy absorbing foam, like the Davisorb bumper system. The total consumption for RIM elastomers for Toyota bumpers is estimated to be 500 tons per month or over 5,000 tons in 1981.

The Davisorb type bumper systems are also applied to Mitsubishi cars, as shown in Table IV. The Lancer EX, whose bumper is an exception to Mitsubishi cars, is equipped with cosmetic bumpers. Similar cosmetic bumpers are used also in some of the Nissan (Datsun) cars.

Another bumper system composed of a bumper cover and an LDPE honeycomb, like the Guide Flex bumper system, is used for the Bluebird and the Leopard. A combination of bumper cover and hydraulic shock absorber is used in the Datsun 280 ZX (Fairlady 280 ZX). A similar bumper system is also applied to a Mazda car, the Savanna

Table II.  Types of Bumper Systems by Make

| TYPE | MAKE |
|---|---|
| 1. RIM COVER + RIM FOAM | TOYOTA, MITSUBISHI |
| 2. RIM COVER + STEEL BEAM + HYDRAULIC SYSTEM | NISSAN, MAZDA |
| 3. RIM COVER + PLASTIC HONEYCOMB | NISSAN |
| 4. RIM COVER + STEEL BEAM | MITSUBISHI, NISSAN |

Table III.   RIM Bumpers by Make (1981 Model) (1)

| MAKE | CAR NAME | BUMPER SYSTEM | PRODUCTION |
|---|---|---|---|
| TOYOTA | CELICA | COVER/FOAM | 500 T/M |
| | CROWN | COVER/FOAM | |
| | MARK II | COVER/FOAM | |
| | CRESSIDE | COVER/FOAM | |
| | CHASER | COVER/FOAM | |
| | CRESTA | COVER/FOAM | |
| | CORONA | COVER/FOAM | |
| | CARINA | COVER/FOAM | |
| | CAMRY | COVER/FOAM | |
| | COROLLA | COVER/FOAM | |
| | SPRINTER | COVER/FOAM | |
| | TURCEL | COVER/FOAM | |
| | CORSA | COVER/FOAM | |
| | STARLET | COVER/FOAM | |

Table IV.   RIM Bumpers by Make (1981 Model) (2)

| MAKE | CAR NAME | BUMPER SYSTEM | PRODUCTION |
|---|---|---|---|
| MITSUBISHI | GALANT SIGMA | COVER/FOAM | 40 T/M |
| | ETERNA SIGMA | COVER/FOAM | |
| | GALANT LAMBDA | COVER/FOAM | |
| | ETERNA LAMBDA | COVER/FOAM | |
| | PLYMOUTH SAPPORO | COVER/FOAM | |
| | DUDGE CHALLENGER | COVER/FOAM | |
| | LANCER EX | COVER | |
| NISSAN | CEDRIC | COVER | 130 T/M |
| | GLORIA | COVER | |
| | GAZELLE | COVER | |
| | SILVIA | COVER | |
| | BLUEBIRD | COVER/HONEYCOMB | |
| | LEOPARD | COVER/HONEYCOMB | |
| | FAIRLADY Z | COVER/HYDRAULIC | |
| MAZDA | SAVANNA RX-7 | COVER/HYDRAULIC | 20 T/M |

RX-7, 1981 model.

Not all bumpers are colored.  Colored bumpers are usually paint-ed by using polyurethane coatings at an off-line temperature of 100-120°C.  Examples of colored bumpers are:  Mitsubishi - Galant Sigma and Galant Lambda; Toyota - Crown, Celica XX, Mark II and, in part, Cresta and Chaser; Mazda - Savanna RX-7.

The production of cars equipped with RIM bumpers in Japan in 1979 is summarized in Table V.  It is interesting to note that the percentage of Toyota cars equipped with RIM bumpers amounted to about 44% of all Toyota cars, and its figures for 1981 are estimated to in-crease by more than 50%.

Table V.  Production of RIM-Bumper Cars
(Source:  Sekiyu Kagakukogyo Kenkyusho)

| MAKE | PRODUCTION | PERCENT |
|------|-----------|---------|
| TOYOTA | 920,000 | CA, 44 |
| NISSAN | 140,000 | CA, 8 |
| MITSUBISHI | 105,000 | CA, 20 |

Examples of down-weighting by means of RIM bumpers in Toyota
cars and Nissan cars are as follows:  The old steel bumper in Toyota
cars, including mechanical shock-absorbers, had a weight of 40 kg per
car; a new RIM bumper has a weight of 30 kg.  Therefore a weight reduction of 25% was achieved.  In the case of the Nissan Bluebird, an
old steel bumper, excluding mechanical shock-absorber, had a weight
of 15.4 kg, and a new bumper composed of bumper cover and honeycomb
core has a weight of 12.2 kg.  In this instance a weight reduction
of 21% and a cost reduction of 26% was achieved.

## Exterior Car Parts

In addition to the use of RIM elastomers for bumpers, the use
of RIM elastomers for exterior car parts is increasing, as shown in
Table VI.  Sight shields are produced by using one of three kinds
of elastomers, i.e., olefinic elastomers, EPDM, or urethane elastomers.  Polyurethane RIM splash guards were employed by Toyota for
the 1977 model Celica, and are being used extensively for other
models.  Lamp housings are made of high impact polypropylene, reinforced polypropylene or SMC.  Recently, however, RIM lamp housings
have been used in Isuzu cars for the first time in Japan since 1980.
Air spoilers are new candidates for automotive RIM, but they are
still under discussion.  Air deflectors are produced by integral
skin polyurethane foam technique, and are expected to be replaced
by RIM polyurethane in the near future.

Some examples of semi-flexible polyurethane RIM foam for automotive uses are shown in Table VII.  Until recently semi-flexible

Table VI.  Automotive RIM Elastomer

SIGHT SHIELDS

SPLASH GUARDS

LAMP HOUSINGS

AIR SPOILERS

AIR DEFLECTORS

Table VII.  Semi-Flexible PU RIM Foam for Automotive Use

1. STEERING WHEELS
2. DASH BOARDS
3. HEAD RESTRAINTS
4. HORN PADS
5. CONSOLE BOX LIDS
6. DOOR TRIMS
7. TRUNK MATS

foam had been widely used in interior parts for safety, e.g., steering wheels, dash boards, etc., by using a low pressure foam-in-place process.  Some of this, however, has recently been replaced by the RIM process because of its improved processability and productivity.

Polyurethane steering wheels have been employed in most Japanese cars.  Total consumption of foam for steering wheels is estimated to be 1,000 tons per year.

The in-mold coating technique combined with the RIM process is very advantageous for producing interior car parts.  In 1979 the meter panel pads for the Celica-Supra were produced by using this technique.  In 1980 the technique was extended by Toyota to produce full instrument panels for Cresta.  Console boxes are another possible application for the RIM process.  Datsun 280 ZX (Fairlady 280 ZX) is equipped with a RIM integral skin console box lid.  Other applications such as horn pads, head restraints, door trims, etc., are expected to develop new RIM markets.  At the same time, however, it is expected that low cost commodity cars will continue for some time to use cheaper plastic materials in place of polyurethanes.  In general, cost reduction has been a lasting requirement in the Japanese car industry, and interior car safety can be achieved by the use of seat belts without the use of shock-absorbing interior parts.  Accordingly, polyurethane interior parts will preferably be used in high grade cars.

NON-AUTOMOTIVE USE OF RIM POLYURETHANES

The application examples of rigid polyurethane RIM foam are shown in Table VIII.  Structural polyurethane foam was first developed by Bayer AG in 1964, and remarkable applications have been reported.  In Japan, however, the foam is still in the development stage, presumably because of the differences in market demand.  The total consumption of rigid polyurethane RIM foam in 1979 was estimated to be in the range of 1,000 - 2,000 tons.  The major application of rigid polyurethane RIM foam lies in sports articles, especi-

Table VIII.  Rigid PU RIM Foam

SPORTS ARTICLES – SKI CORES, RACKET FRAMES, RUDDERS.

INTERIOR ARTICLES – CHAIR SHELLS, SANITARY WARES,
WATER TANKS FOR TOILET,
PICTURE FRAMES, SIMULATED WOOD.

APPLIANCE – TV CABINETS, AIR CONDITIONER HOUSINGS.
DRAIN PANS FOR REFRIGERATOR, PROJECTOR CABINETS
SPEAKER BOXES.

OTHERS – FILTER FRAMES, WINDOW FRAMES, BARBER CHAIRS.

ally in ski cores.  Interior articles such as chair shells have also been a major application of molded rigid urethane foam.  An advantage of the RIM process lies in cheaper tooling and re-tooling costs, which significantly reduces the production costs in comparison with steel plate.  The process is therefore preferred for the production of articles in small quantity.  Appliance uses, e.g., TV cabinets, air conditioner housings, drain pans for refrigerators, projector cabinets and speaker boxes are promising markets for RIM products.  Other applications, e.g., filter frames, window frames, barber chairs, beauty parlor chairs, dental chairs, etc., have been considered to be promising markets for RIM products.  In particular, barber chairs have been manufactured extensively on a world-wide basis by Takara Chair Manufacturing Co., Ltd.

The RIM ski core production in Japan in 1979 is shown in Table IX.  The annual production of ski cores in 1979 was 300 to 400 tons. The potential market of polyurethane ski cores is estimated to be 600 to 900 tons per year in the near future.  Until recently ski cores have been produced by using low pressure machines.  Very recently, however, the RIM process has been used in the production of ski cores by Swallow Ski Co., Ltd., and Nippon Gakki Co., Ltd.  Other ski makers, e.g., Mizuno Co., Ltd., and Awoyama Ski Co., Ltd., are planning to use the RIM process for producing ski cores.

Table IX.  RIM Ski Core Production (1979)
(Source:  Sekiyu Kagaku Kenkyusho)

| COMPANY | RIM SKI CORE, T/Y |
| --- | --- |
| NIPPON GAKKI | 120 |
| SWALLOW SKI | 130 |
| MIZUNO | 33 |
| AOMORI SKI | 20 |
| OTHERS | 40 |
| TOTAL | 300 – 400 |

AUTOMOTIVE AND NON-AUTOMOTIVE USE OF RRIM POLYURETHANES

At the present time there are few RRIM products on the Japanese market, in contrast to the extensive efforts that have been focussed on RRIM for cars in both the U.S.A. and Europe.  There are many reason for the slow commercialization of RRIM in Japan.  The first of these is the cost comparison with steel.  In Japan steel has been the most economical structural material because of its improved productivity and high quality.  High quality steel makes it possible to use much thinner steel sheet, i.e., the thickness of conventional steel sheet used for cars has been 0.7 mm.  In recent years high strength steel has also been used for some exterior body panels, which makes it possible to reduce the thickness from 0.7 mm to 0.65 mm, or a 7% reduction in both the thickness and cost.  Therefore, at the present time the cost of RRIM fenders in Japan is estimated to be higher than steel fenders.

Secondly is the comparison between insurance premium and repair cost of steel fenders.  Fenders made of RRIM may have non-damageability in the case of low speed impact, but high speed impact may result in replacement of RRIM fenders.  The damaged steel fenders can be repaired at a lower cost.  The cost of replacement versus repair is an area that auto manufacturers and insurance companies will want to examine.

Thirdly, the processability of steel fenders is much more efficient than that of RRIM fenders.  The stamping cycle of steel fenders is only a few seconds, while the molding cycle of RRIM fenders is more than one minute.

Fourthly, the corrosion warranty of RRIM fenders is recognized to be an advantage, but in Japan salt is not used on the roads in the snow season; therefore, few requirements are given for the corrosion warranty of car body panels.

Fifthly, at the present time on-line painting of RRIM fenders is not possible.

Due to the uncertainty of the RRIM development, a prudent attitude is the motto of Japanese car manufacturers in the use of RRIM fenders and other exterior body panels.  Still they earnestly desire that some innovative technologies will develop which will overcome the said disadvantages, and some Japanese car manufacturers, in fact, have developed the RRIM process to a certain extent.  They look forward to successful commercialization of RRIM fenders in the U.S.A. and in Europe.  Possible future applications of RRIM (reinforced RIM), including automotive and non-automotive uses, are shown in Table X.

Table X.   Possible Use of RRIM

"ARTICLES IN SMALL QUANTITY AND WIDE VARIETY OF PRODUCTION"

AUTOMOTIVE USE : EXTERIOR BODY PANELS
                          – DOOR PANELS FOR TRUCKS
                          – REAR DOOR FOR STATION WAGONS

NON-AUTOMOTIVE USE : HOUSINGS FOR   – COMPUTERS
                                    – COPYING MACHINES
                                    – ELECTRIC EQUIPMENTS
                 : STREET FURNITURE
                 : FLOWER POTS

As mentioned before, one of the advantages of RRIM lies in the economical tooling and retooling costs.   RRIM thus is considered to be well suited for articles in small quantity and wide variety.   In the case of automotive use, exterior body panels such as door panels for trucks and rear doors for station wagons may be possible candidates in the initial stage of RRIM production, because they are now in small production compared to cars.

In the case of non-automotive uses of RRIM, various housings for computers, copying machines, electric equipment, street furniture and flower pots are identified as possible applications.

NON-URETHANE RRIM

In addition to the competition of urethane RRIM with steel, other plastics such as nylon 6, epoxides, polyesters and vinyl esters are considered to be competitive materials with polyurethane RRIM. However, development efforts for non-urethane RRIM have not been reported in Japan as yet.

RIM MACHINES

The RIM machines available in Japan are listed in Table XI.   At the present time it is estimated that there are about 30 RIM machines in operation in Japan.   Admiral and the other eight machine manufacturers supply their machines from the U.S.A. and Europe.   Maruka Kakoki Co., Ltd., Niigata Tekko Co., Ltd./Asahi Glass Co., Ltd., and Toho Kikai Co., Ltd., have developed RIM machines by their own technologies.

RAW MATERIALS FOR RIM/RRIM

The TDI producers in Japan, along with the production capacities

Table XI.   RIM Machines

| MANUFACTURER | DEALER OR LICENSEE |
|---|---|
| ADMIRAL EQUIPMENT | ADMIRAL CO. OF JAPAN |
| AFROS-CANNON | MITSU-NISSO URETHANE |
| CINCINNATI MILACRON | NIPPON CINCINNATI MILACRON |
| DESMA | TOSHIBA KIKAI |
| ELASTOGRAN | POLYURETHANE ENGINEERING |
| HENNECKE | SUMITOMO BAYER URETHANE |
| KRAUS-MAFFEI | K. BRASCH & CO. LTD. |
| MARTIN SWEETS | TOYO KASEI |
| MARUKA KAKOKI | ←——— |
| NIIGATA TEKKO / ASAHI GLASS | ←——— |
| TOHO KIKAI | ←——— |
| VIKING ENGINEERING | CHORI |

and technologies employed, are listed in Table XII.   In Table XIII the MDI producers are shown.   Due to the increasing demand for RIM products, expansion of the production capacity of MDI is projected by all MDI producers.   In 1982, the total production capacity of MDI in Japan will be about 108,000 tons.

All of the said polyisocyanate manufacturing plants are employing the conventional method, or the so-called phosgen method.   Recently, however, a phosgen-free method for TDI production has been announced by Mitsui Toatsu Chemical and Mitsubishi Chemical (see Figure 9).   The method is composed of a two-step process, i.e., carbonylation of dinitrotoluene, followed by thermal dissociation of the diurethane.   The development of the phosgen-free technology was first announced by Mitsui-Toatsu Chemical Co., Ltd., in February, 1978.   Very recently Mitsubishi Chemical Ind., Ltd., announced that a pilot plant having a production capacity of 500 tons per year will be forthcoming in mid-1981.   Some of the advantages of the phosgen-free method are:   (a) no use of phosgen and no need for the electrolyses plant and (b) lower energy cost and lower material cost.

A similar technology has reportedly been developed by Arco

Table XII.   TDI Producers in Japan (T/Y) (1981)
(Source:   Communication)

| COMPANY | CAPACITY | TECHNOLOGY |
|---|---|---|
| MITSUBISHI CHEMICAL | 9,600 | OWN TECHNOLOGY |
| MITSUI TOATSU CHEM. | 25,000 | DU PONT |
| NIPPON POLYURETHANE | 13,500 | BAYER/NPU |
| SUMITOMO BAYER | 13,000 | SUMITOMO CHEM./BAYER |
| TAKEDA CHEMICAL | 17,000 | ALLIED |

Table XIII.  MDI Producers in Japan (Source:  Communication)

| COMPANY | CAPACITY | | TECHNOLOGY |
|---|---|---|---|
| | 1980 | 1982 | |
| MCI/KASEI UPJOHN | 6,600 | 35,000 | UPJOHN COMPANY |
| MITSUI TOATSU CHEM. | 9,000 | 15,000 | OWN TECHNOLOGY |
| NIPPON POLYURETHANE | 24,000 | 30,000 | NPU/ICI |
| SUMITOMO BAYER URETHANE | 16,000 | 28,000 | BAYER AG. |

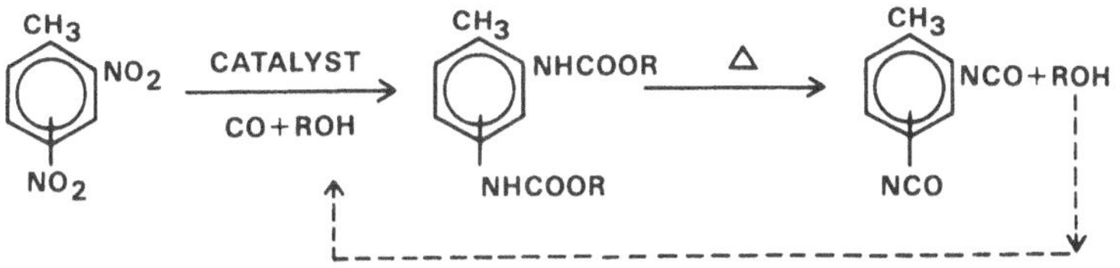

Figure 9.  Phosgen-free TDI method.  Mitsui Toatsu Chemical:  announced in Feb., 1978; Mitsubishi Chemical:  announced in Aug., 1980 (500 T/Y)

Chemical Corporation for producing MDI.  This method (see Figure 10) is composed of the two-step process as seen in the TDI production. However, the condensation reaction of phenyl carbamate with formaldehyde is slower than the reaction of aniline with formaldehyde and therefore a higher yield of MDI with a smaller amount of oligomers

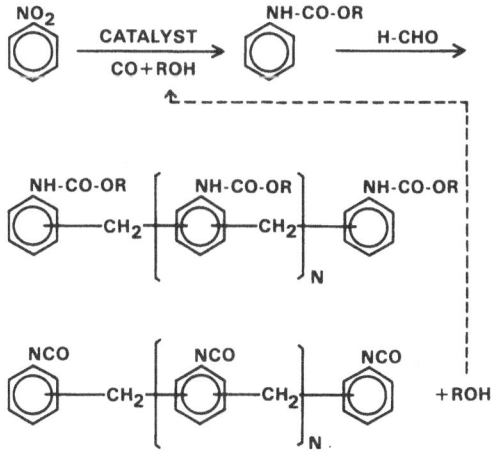

Figure 10.  Phosgen-free MDI method.

is achieved.  This method might be a promising one to produce MDI,
although much effort still is required to establish an industrial
technology.

The conventional polyether polyol producers in Japan are listed
in Tables XIV and XV, but only three companies listed in Table XV are
producing polymer polyols.  Both Mitsui Nisso Urethane Co., Ltd., and
Sanyo Chemical Co., Ltd., are employing grafting reaction of vinyl
monomers in polyether polyols.  Asahi Denka/Miyoshi Oil and Fat Co.,
Ltd., are employing a different method, i.e., a polyether polyol is
reacted with itaconic acid to form a polyether polyester which is
further reacted with a vinyl monomer to yield a grafted polyol.  The
itaconic acid inherently has a double bond and it is believed, therefore, that the grafting reaction is unquestionable.

The PTMEG (polytetramethylene ether glycol) producers in Japan
are listed in Table XVI.  Both Mitsubishi Chemical Ind., Ltd., and
Hodogaya Chemical Ind., Ltd., employ an anionic polymerization reaction of tetrahydrofuran to produce PTMEG.  The chain extender producers in Japan are listed in Table XVII.  The representative chain
extenders for RIM are 1,4-butanediol and ethylene glycol.  1,4-bu-
tanediol is produced by three companies:  Mitsubishi Chemical Ind.,
Ltd., Toyo Soda Co., Ltd., and Daicel Chemical Co., Ltd.  Each uses
its own technology.  Mitsubishi Chemical Ind., Ltd., has commercial-
ized the product by using the method shown in Figure 11.  Acetoxyl-

Table XIV.  Polyether Polyol Producers in Japan (1980) (Source:  JURA)

| | |
|---|---|
| A S KASEI | MITSUI NISSO URETHANE |
| ASHI OLIN | NIPPON OIL AND FATS |
| DAINIPPON INK AND CHEMICALS | SANYO CHEMICAL |
| DAI-ICHI KOGYO SEIYAKU | SUMITOMO BAYER URETHANE |
| HODOGAYA CHEMICAL | TAKEDA CHEMICAL IND. |
| MITSUBISHI CHEMICAL | TOHO CHIBA CHEMICAL IND. |

Table XV.   Polymer Polyol Producers in Japan   (Source:
Communication with the Companies)

| COMPANY | CAPACITY | TECHNOLOGY |
|---|---|---|
| ASAHI DENKA-MIYOSHI OIL & FAT | 4,000 | OWN TECHNOLOGY |
| MITSUI NISSO URETHANE | 10,000 | UCC |
| SANYO CHEMICAL | 6,000 | BAYER |

Table XVI.  PTMEG Producers in Japan (1980)

| COMPANY | CAPACITY (T/Y) | SOURCE |
|---|---|---|
| MITSUBISHI CHEMICAL | 2,500 | NEWSPAPERS |
| HODOGAYA CHEMICAL | 1,500 | ESTIMATED |

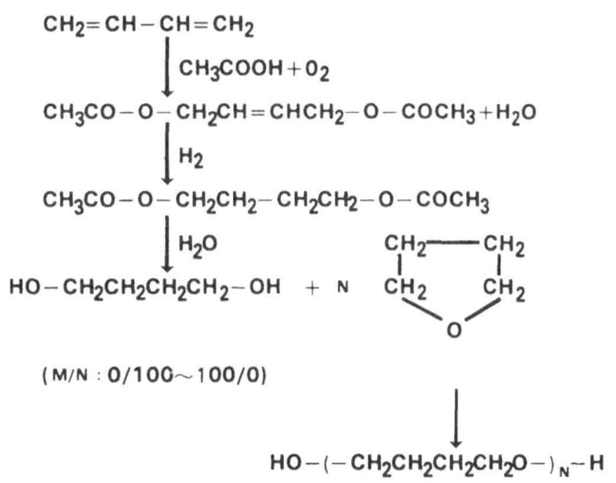

$$CH_2=CH-CH=CH_2$$

$$\downarrow CH_3COOH+O_2$$

$$CH_3CO-O-CH_2CH=CHCH_2-O-COCH_3+H_2O$$

$$\downarrow H_2$$

$$CH_3CO-O-CH_2CH_2-CH_2CH_2-O-COCH_3$$

$$\downarrow H_2O$$

$$HO-CH_2CH_2CH_2CH_2-OH \quad + \quad N$$

$$( M/N : 0/100 \sim 100/0 )$$

$$\downarrow$$

$$HO-(-CH_2CH_2CH_2CH_2O-)_N-H$$

Figure 11.  1,4-Butanediol & PTMEG (Mitsubishi Method)
           (Source:  publications)

ation of butadiene, followed by hydrogenation and hydrolysis, leads
to the simultaneous formation of both 1,4-butanediol and tetrahydrofuran at any ratios depending upon the reaction conditions.  The tetrahydrofuran can be further polymerized to produce PTMEG.  The Toyo
method (Figure 12) is composed of chlorination of butadiene, followed
by hydrogenation and hydrolysis of the intermediate product.  The
Daicel method (Figure 13) is composed of oxo reaction of allyl alco-
hol which is derived from propylene oxide, followed by hydrogenation
of the intermediate product.

Ethylene glycol is produced in large quantities by four of the
companies listed in Table XVII, and most of the product is consumed
for the production of polyester fibers.  Ethylene glycol is produced
from ethylene oxide via direct oxidation of ethylene.  The chloro-
hydrin method for producing ethylene oxide is no longer used in Japan.

THE FUTURE OF RIM/RRIM IN JAPAN

It is believed that automotive RIM elastomers will be increasingly used in bumpers and probably in fascias and other exterior

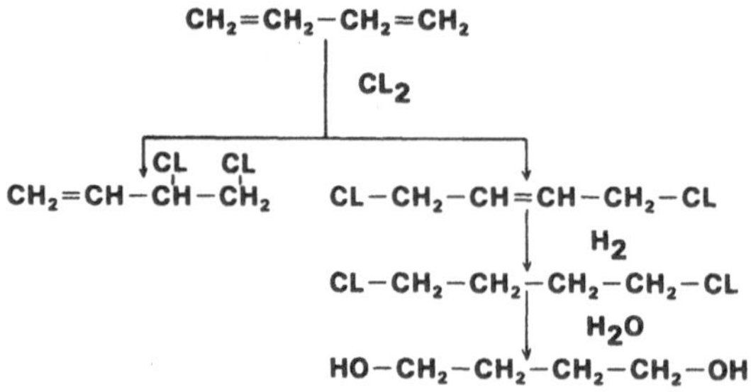

$$CH_2=CH_2-CH_2=CH_2$$

$$CL_2$$

$$CH_2=CH-CH-CH_2 \qquad CL-CH_2-CH=CH-CH_2-CL$$
$$\quad CL \quad CL$$

$$H_2$$

$$CL-CH_2-CH_2-CH_2-CH_2-CL$$

$$H_2O$$

$$HO-CH_2-CH_2-CH_2-CH_2-OH$$

Figure 12.  1,4-Butanediol (Toyo Soda method) (Source:  publications)

$$CH_2-CH-CH_3$$
$$\quad O$$

$$CH_2=CH_2-CH_2-OH$$

$$CO+H_2$$

$$OHC-CH_2-CH_2-CH_2-OH$$

$$H_2$$

$$HO-CH_2-CH_2-CH_2-CH_2-OH$$

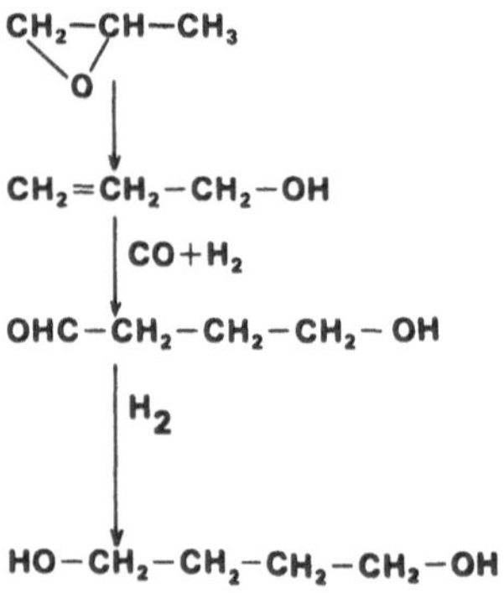

Figure 13.  1,4-Butanediol (Daicel method) (Source:  publications)


parts.  The possible use of RIM for bicycle tires has been in the
discussion stage, while RIM bicycle chains are reported to appear on
the Japanese market very soon.  The production of tractor tires has
not been attempted as yet; tractor tires and other off-road tires
have been imported from Austria.

     The production of RIM passenger car tires seem to have a tech-
nical barrier to overcome which is in low thermal stability of urethane elastomers.  The Bridgestone Tire Co., Ltd., recently developed
a urethane-based liquid rubber having superior physical properties
which are equal to those of natural rubber.  The development is based
on a new finding that a filler having a particle size of ca. 1/3 of
the distance between adjoining urethane linkages increases mechanical
strengths of polyurethane elastomers up to those of natural rubber.
For example, a PTMEG-based prepolymer, MOCA and silica powder having
the said particle size are mixed and cast into a mold to form a shaped
product.  Another example is that polybutadiene-based prepolymer, MOCA
and carbon black having the said particle size are blended and cast

Table XVII.   Chain Extender Producers in Japan (1981, projected)
              (Source:  MITI)

| CHAIN EXTENDER | PRODUCER | CAPACITY |
|---|---|---|
| 1,4-BUTANE DIOL | MITSUBISHI CHEMICAL | 15,000 |
| | TOYO SODA | 6,000 |
| | DAICEL CHEMICAL IND. | 5,000 |
| ETHYLENE GLYCOL | JAPAN CATALYTIC CHEM. | 225,000 |
| | MITSUBISHI PETROCHEM. | 185,000 |
| | MITSUI PETROCHEMICAL | 145,000 |
| | NISSO PETROCHEMICAL | 128,000 |

(SOURCE : MITI)

into a mold to make a molded product.  The filled urethane elastomers have two times higher tensile strength, three times higher elongation and five times higher tear strength than the existing liquid rubber-based elastomers.

RIM elastomers can find an application in the railroad industry. Shin-Kansen, the so-called "Bullet Train" which travels at a speed of 200 km/h or 125 miles per hour, gives serious noise pollution in the urban areas.  RIM elastomers were applied to cross-ties for Tohoku Shin-Kansen which was recently constructed in the northern part of Japan.  The cross-ties made of concrete were lapped with urethane elastomers by using the RIM process at the railroad construction fields.  This application exhibited remarkable noise-damping effects.  Therefore RIM cross-ties could be extensively used for the railroads and subways in urban areas in the near future.

RRIM might be practically used as exterior body panels in Japanese cars if some innovative technologies are developed that could overcome the said disadvantages of RRIM in comparison with steel. The following new innovative technologies are needed for RRIM:

(1) Improvement of the physical properties:  As discussed previously, improvements in high temperature resistance of RRIM are necessary in order to make on-line painting possible.  However, the urethane linkage inherently is not thermally stable.  Thus technical improvement in this area may be in the modification of polymer structure by incorporating high temperature resistant linkages.  The incorporation of a thermally stable linkage, e.g., isocyanurate linkage, was attempted but it resulted in inferior properties at low temperatures.  An innovative chemistry in isocyanate-based polymers is long awaited.

(2) Low temperature coating:  As stated above, the most desirable innovation awaited for RRIM fender production lies in the development of high temperature-resistant polymer structures.  If that is not

possible, a counter measure for RRIM fender production is the use of low temperature cure coatings produced at a competitive price with the existing coatings. Low temperature coatings which can meet the requirements are a key factor for the RRIM fender production.

(3) Internal releasing agents: This is not an essential part of RRIM chemistry, but it is very important from the standpoint of the productivity of RRIM fenders. The use of an internal releasing agent, therefore, is another key factor in the RRIM fender production on the basis of the existing technology.

(4) Waste disposal: This item relates to the environmental problems in both the production of RIM products and the disposal of cars. Incineration of wastes may be a cause of air pollution and loss of resources. Land fill is a possible method, but in Japan land fill is under a restriction. In addition the method is not desirable in our resource conservation policy. Recycling of raw materials is another method, but RIM products have different compositions and the raw materials recovered can be used only in limited products. Re-use of RIM and RRIM products to other articles may result in limited applications, so the method still cannot be used widely.

(5) Raw material costs: The price increases of RIM raw materials, especially polyisocyanates and polyols, is a continuing problem in the competition with other materials such as other plastics, high tensile steel, and aluminum. Thus innovative technologies for producing polyurethane RIM raw materials by using energy-saving processes are subjects for the further development of polyurethane RIM products.

In conclusion, the use of regular RIM products for various vehicles will definitely be increased in Japan. Additional successful use of RIM depends upon improved technology.


References

1.  H. Kaneko, Sekiyu Kagaku Kogyo Kenkyusho (Petrochem. Ind. Investigation, Inc.), 1-19-6, Nishi-shinbashi, Minato-ku, Tokyo, 105, Japan.
2.  JURA (Japan Urethane Raw Materials Assoc.), No. 1 Mori-Bldg. 1-12-1, Nishi-shinbashi, Minato-ku, Tokyo, 105, Japan.
3.  JAMA (Japan Automobile Manufacturers Assoc., Inc.), Otemachi Bldg. 6-1, Otemachi-1-chome, Chiyoda-ku, Tokyo, 100, Japan. Washington office: 1050 17th St. N.W., Washington, D.C. 20036
4.  H. Igami, Polymer Preprint (Japan) 27, 4, 680 (1978).

STUDIES OF THE FORMATION AND PROPERTIES OF POLYURETHANES

SUITABLE FOR REACTION INJECTION MOULDING

J. L. Stanford, R. F. T. Stepto and R. H. Still

Department of Polymer and Fibre Science
The University of Manchester Institute of
    Science and Technology
Manchester, M60 1QD, England

INTRODUCTION

From fundamental and technological points of view RIM is a complex process.[1] Fast reactions are required to form, in many applications, network materials and the reactions must by synchronised with the process so that the mould is filled with liquid reacting material, the material becoming solid at complete filling. The rates of reaction occurring between the two liquid reactant feeds which are mixed prior to mould filling must be well controlled, as must be the molecular growth process leading to the final, solid material. The important point in the polymerization is the gel point, which in terms of reactant flow defines the point of infinite viscosity.[2,3]

Polyurethane-forming systems have been of paramount importance in the development of RIM, where reactant streams are low viscosity prepolymers or monomers which react to yield networks. The urethane, network-forming reactions may be performed at ambient temperatures and their rates catalytically controlled. In addition volatile evolution may be obtained, if desired, by the deliberate incorporation of blowing agents. Large and complex mouldings can be produced in non-cellular and microcellular, filled and unfilled materials, thereby increasing the structural and engineering applications of polymers.

The properties of linear polymers depend principally on their chain structures and molar masses. The PU materials formed by RIM are often based on linear, segmented urethanes to give networks whose junction points are essentially due to hydrogen-bonding between the hard blocks. Additional, covalent junction points may be introduced by the use of polyfunctional reactants, thus avoiding the fall-off

31

in physical properties at elevated temperatures which occurs with
hydrogen-bonded, thermoplastic elastomers. In this context, the
present work has focussed on the formation and properties of cova-
lently-linked network materials which may be formed by RIM. One-
phase PU networks of relatively well-defined structures are considered so that materials properties can be more directly related to
reactant structures and reaction conditions than with segmented materials. Although, in terms of commercial RIM formulations, the present studies deal with model network-forming reaction systems, some
of the materials produced have physical and mechanical properties
which make them suitable for a wide range of applications.

Network materials have properties which are dependent on (a)
reactant molar mass, (b) reactant functionality, and (c) structural
features of the reactants (affecting chain flexibility). The expected molar mass $M_c^o$, or chain length, between junction points and
the junction-point functionality, f, are defined by (a) and (b) respectively. Furthermore, side reactions (allophanate, biuret, etc.
in PU formation) and intramolecular reactions occurring during network formation yield structural modifications which also affect the
physical properties of the formed materials. Although side reactions
often occur and can be desirable in industrial formulations, it is
possible to prepare polyurethanes with the negligible occurrence of
side reaction.[4-7] However, intramolecular reaction is an integral
part of the random (or condensation) polymerization process.[4,6,8-10]
In an otherwise ideal network-forming system, the amount of intramolecular (loop-forming) reaction occurring en route to gelation is
influenced by (a), (b) and (c).[5-7,10-21] Such intramolecular reaction results in structural defects in the final network, and these
defects have been shown to affect physical properties.[14-18,21] For
a given reaction system, the numbers of such defects may be varied
by carrying out reactions at different initial dilutions in an inert
solvent.

The present account gives a review of our recent and published
studies on PU systems with possible applications to RIM. Batch polymerizations of model systems have been used to illustrate the dependence of materials properties on the parameters (a), (b) and (c).
Thus, purified hexamethylene diisocyanate (HDI) and 4,4'-diphenyl-
methane diisocyanate (MDI) have been reacted with commercially available polyoxypropylene (POP) triols of various molar masses. In addition, to explore the range of non-cellular materials obtainable,
especially synthesized tetrols have been employed.[7,16,20]

Comparative studies have also been made between filled and unfilled systems. The former are relevant to RRIM, in which reinforcing filler is introduced into one or both of the reacting systems
to produce primarily materials of higher stiffness and reduced thermal expansion. The effects of such fillers on materials properties

depend principally on fibre orientation, aspect ratio and volume fraction. In addition, the interaction between fibre and matrix and the properties of the unfilled matrix itself play an important role. Thus, the work has used glass fibres and has included the effects of filler concentration and fibre length on the physical and mechanical properties of model PU matrices.[18,22]

In all the reaction systems, polyols of known functionality and molar mass have been used and reactions have been carried out using stoichiometric amounts of reactants. Thus the chemical structure of the network which is formed from given reactants is known, and in particular the molar mass between junction points of the perfect network, $M_c^o$, is defined. $M_c^o$ is an important parameter for comparative descriptions of the effects of loop formation on moduli of actual networks. The definition of $M_c^o$ for HDI reacting with a POP tetrol based on pentaerythritol (PE) is given in Figure 1 (a) and (b). Also in Figure 1 (c), a quantity, $\nu$, is defined which is the number of bonds in the chain forming the smallest ring structure. Comparing Figure 1 (b) and (c), it can be seen that $\nu$ is closely related to $M_c^o$. In fact, apart from a possible difference of one or two bonds resulting from the reaction of pairs of groups, $\nu$ is equal to the chain length corresponding to molar mass $M_c^o$.

EXPERIMENTAL INVESTIGATIONS

These fall into three groups:
    (i) Correlations between gel points and materials properties.
    (ii) Gelation and pre-gel intramolecular reaction studies.
    (iii) Correlations between reactant characteristics and materials properties.

Studies (i) have been made on systems where the reaction rates were slow enough (see Table III later) to allow sampling of the reaction mixture followed by chemical analysis to evaluate extents of reaction of -NCO and -OH groups, $p_a$ and $p_b$, respectively. The physical properties of the materials formed at complete reaction have been investigated and related to the products of the extents of reaction at gelation $(p_a p_b)_c$ ( $= \alpha_c$), of the reaction mixtures.[16-18,21]

Such investigations have indicated that $\alpha_c$ is an important parameter for characterizing physical properties, and the investigations (ii) have shown in turn how $\alpha_c$ and pre-gel intramolecular reaction are related to reactant structure and molar mass.[5-7,10-21]

The studies (iii) have investigated the physical properties of filled and unfilled polyurethane materials. The polyurethane matrices in these materials have been formed from MDI reacting with a POP triol in admixture with trimethylol propane (TMP). The fillers used have included hammer-milled and chopped glass fibres. $T_g$, shear

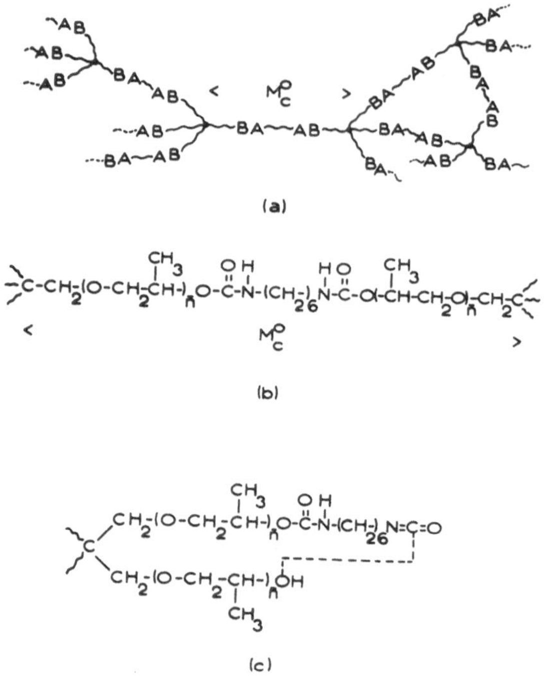

(a)

(b)

(c)

Figure 1.    (a) Part of tetrafunctional network structure formed from
             an $RA_2 + RB_4$ polymerization corresponding to $M_c^o$, the molar
             mass between junction points of the perfect network.
             (b) Detail of chain structure defining $M_c^o$ for HDI reac-
             ting with an oxypropylated pentaerythritol (OPPE). $\bar{n}$ is
             the number-average degree of polymerization of each arm
             of the tetrol with respect to propylene oxide units.
             (c) Part of the chain structure defining $\nu$, the number
             of bonds in the chain forming the smallest ring struc-
             ture, for the reaction system in (b).

modulus, dynamic-mechanical, tensile and impact properties have been
measured and interpreted, where possible, in terms of the reactant
parameters (a), (b) and (c), described previously, and also in terms
of filler content and characteristics.[21,22]  The reaction rates were
such that gel times ranging from seconds to minutes were encountered,
without the use of added catalyst.  In the context of RIM and RRIM,
the faster reacting systems are the more relevant.

(i) Correlations Between Gel Points and Materials Properties

    Gel points of various systems have been determined, as described
in the literature,[5,11]  including those to be considered in detail
in the present paper, namely, HDI and MDI reacting with POP triols

and tetrols at 80°C in bulk and at various dilutions in nitrobenzene as solvent.[16-18,21] Aliquots from reaction mixtures have also been allowed to proceed to complete reaction in moulds to form networks. Stress-strain behaviour, static and dynamic moduli and relaxational behaviour, including Tg, of the networks were evaluated after removal of any sol-fraction.

Shear Modulus. Figure 2 illustrates some of the correlations obtained between gel points and shear modulus.

In the ordinate in Figure 2, $M_c$ is the molar mass between junction points, evaluated from values of rubbery shear moduli, G, according to the equation:[23]

$$G = ART\rho \ \phi_2^{1/3} \ (V_u/V_F)^{2/3}/M_c \tag{1}$$

$\rho$ is the density of the dry network, $\phi_2$ is the volume fraction of solvent present in the swollen network, $V_u$ the volume of the dry, unstrained network, and $V_F$ the volume of the network at formation. G was determined from small-strain uniaxial compression and torsion pendulum (1 Hz) measurements for dry ($\phi = 1$) and swollen networks. A, having the value (1 - 2/f) for phantom networks and 1 for networks showing affine behaviour,[24] can be put equal to 1, in view of the small strain measurements used.[25] Thus, the ordinate in Figure 2 is an inverse scale of values of G of dry networks, relative to those for the perfect networks, with $M_c/M_c^o = 1$ corresponding to the moduli when, for all systems, $M_c = M_c^o$.

The abscissa in Figure 2, $p_{r,c}$, is the fraction of groups which have reacted intramolecularly at the gel point, and is defined by the equation:

$$p_{r,c} = \alpha_c^{1/2} - (\alpha_c^o)^{1/2}, \tag{2}$$

where $\alpha_c^o(= 1/f-1))$ is the value of $\alpha_c$ in the absence of intramolecular reaction.[26] Figure 2 contains results for tri- and tetrafunctional networks, and for both types $p_{r,c} = 0$ defines the perfect gelling system. The ranges of values of $p_{r,c}$ for various systems arise from the different initial dilutions of the reaction mixtures used to prepare the networks, and depend on the polyol functionality.

Results from five series of networks are shown in Figure 2; systems 1 and 2 refer to HDI/POP triol networks with two values of $M_c^o$, system 3 to MDI/POP triol networks, and systems 4 and 5 to HDI/POP tetrol networks with two values of $M_c^o$. In all cases, $M_c/M_c^o > 1$ and tends to 1 as $p_{r,c} \longrightarrow 0$. Thus, only in the limit of a perfect gelling system is a perfect network achieved. The pre-gel intramolecular reaction, which causes $\alpha_c$ to exceed 1/(f-1) in value, also produces elastically ineffective loops which have marked effects on the moduli of the dry networks. The different slopes of the pairs

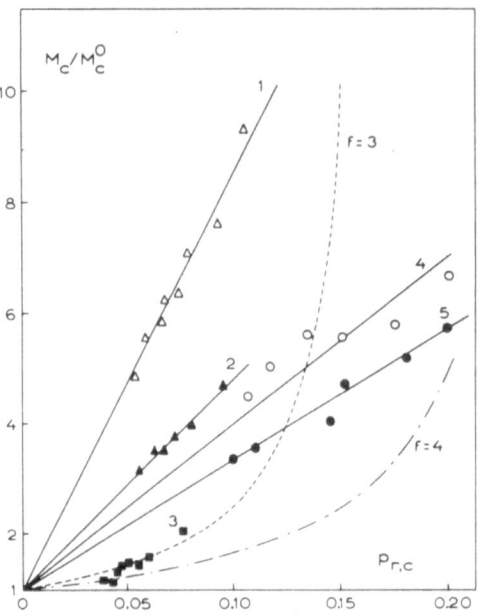

Figure 2.   Molar mass between elastically effective junction points
($M_c$) relative to that for the perfect network ($M_c^O$) versus
extent of intramolecular reaction at gelation ($p_{r,c}$) for
PU networks.
———— Lines through experimental points (systems 1, 2, 4
and 5); ——— and -·-·- calculated curve for trifunctional
and tetrafunctional networks, respectively, assuming that
network defects are introduced by pre-gel intramolecular
reaction and that this leads to only the smallest loops
(those of $\nu$ bonds).
Systems 1 and 2, HDI/POP triols; 3, MDI/POP triols; 4 and
5 HDI/POP tetrols.
System 1, HDI/LHT240 $M_c^O$ = 635 g mol$^{-1}$, $\nu$ = 33; system 2
HDI/LHT112, $M_c^O$ = 1168 g mol$^{-1}$, $\nu$ = 61; system 3, MDI/
LHT240, $M_c^O$ = 705 g mol$^{-1}$, $\nu$ = 30; system 4, HDI/OPPE-NH1,
$M_c^O$ = 500 g mol$^{-1}$, $\nu$ = 29; system 5 HDI/OPPE-NH2, $M_c^O$ = 586
g mol$^{-1}$, $\nu$ = 33.   (LHT240 and LHT112 - oxypropylated
1,2,6 hexane triols).

of lines for similar systems, namely, 1 and 2, and 4 and 5, show that
larger proportions of the loops formed pre-gel are elastically ineffective the lower is the value of $M_c^O$ (or $\nu$) for the reactants.  That
is, system 1 shows higher values of $M_c/M_c^O$ than system 2 for a given
value of $p_{r,c}$, and system 4 shows higher values than system 5.  The
same relative behaviour has also been found for POP triol-based polyester networks.[14]

The broken curves in Figure 2 are the expected values of $M_c/M_c^o$ for trifunctional and tetrafunctional networks, assuming that all the ring structures formed are of the smallest size ($\nu$ bonds), and that only the ring structures formed pre-gel give elastically ineffective loops. Such structures within networks are illustrated in Figure 3. For f = 3, each pair of intramolecularly reacted groups removes two junction points,[14] whereas for f = 4 only one junction point is removed. Simple calculation shows that for f = 3

$$\left(\frac{M_c}{M_c^o}\right)_3 = \frac{1}{(1 - 6p_{r,c})} \, , \tag{3}$$

and for f = 4

$$\left(\frac{M_c}{M_c^o}\right)_4 = \frac{1}{(1 - 4p_{r,c})} \, . \tag{4}$$

The broken curves in Figure 2 are plotted according to equations (3) and (4) and show that intramolecular reaction has a more marked effect on the modulus of trifunctional as compared with tetrafunctional networks. Such relative behaviour is in keeping with that shown by the experimental results, systems 1 and 2 as compared with systems 4 and 5. However, the results for these systems do not follow the calculated curves, indicating that more complicated ring structures, which remove relatively more junction points, are contributing to the elastically ineffective loops. However, the results for system 3 (aromatic diisocyanate/triol-based polyurethanes)

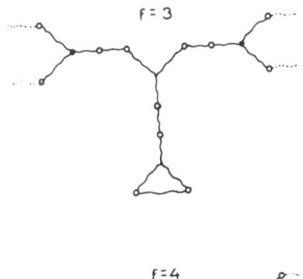

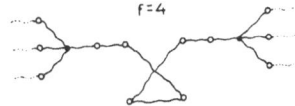

Figure 3.  Ring structures of the smallest size in networks formed from $RA_2 + RB_3$ and $RA_2 + RB_4$ polymerizations.
●-elastically effective junction points.
o pairs of reacted groups (-AB-).

do follow the predictions of equation (3). The simplest interpretation would be that the stiffer MDI/POP chain in this system yields a different distribution of sizes of ring structures, in which rings are predominantly of the smallest size, and, in keeping with equation (3), that all network defects are produced by pre-gel intramolecular reaction. This interpretation is obviously over-simplified as there is no reason a priori why the distribution of ring sizes should be different for system 3, and some elastically ineffective loops must be introduced post-gel. The smaller values of $M_c/M_c^o$ (or higher relative moduli) for system 3 could also result from the stiffer, aromatic chains giving stronger inter-chain interactions.

The functionality of a reaction system and its resulting network affects the scales of both axes in Figure 2. As far as the smallest, pre-gel ring structures are concerned, the effect of functionality on modulus may be removed by plotting $M_c/M_c^o$ normalized to the values expected on the basis of f = 3 and f = 4. Taking f = 4 as a basis, $(M_c/M_c^o)_4$ values remain unchanged and $(M_c/M_c^o)_3$ values become

$$\left(\frac{M_c}{M_c^o}\right)_{3 \to 4} = \left(\frac{M_c}{M_c^o}\right)_3 \cdot \left[\frac{(1 - 6p_{r,c})}{(1 - 4p_{r,c})}\right] \tag{5}$$

In this way, a single theoretical curve results for trifunctional and tetrafunctional systems and the ordinate of the experimental points for f = 3 systems are reduced as

$$\left(M_c/M_c^o\right)_{3 \to 4} < \left(M_c/M_c^o\right)_3$$

The abscissa in Figure 2 has the range 0 to $1 - 2^{-1/2}$ for f = 3 systems and 0 to $1 - 3^{-1/2}$ for f = 4 systems. However, if $p_{r,c}$ is divided by $1 - (\alpha_c^o)^{1/2}$, a scale of 0 to 1 results for both functionalities. Thus, from equation (2), one may define

$$\tilde{p}_{r,c} = \left(\alpha_c^{1/2} - (\alpha_c^o)^{1/2}\right) \Big/ \left(1 - (\alpha_c^o)^{1/2}\right), \tag{6}$$

where $\tilde{p}_{r,c}$ is the fractional extent of intramolecular reaction at gelation on the basis of the total amount possible, namely, $1 - (\alpha_c^o)^{1/2}$.

The rescaling of the two axes in Figure 2 according to equations (5) and (6) gives the plots shown in Figure 4. Approximate agreement with theory still exists for the aromatic system, and $M_c/M_c^o$ for systems 1 and 2, related to systems 4 and 5, show a reduced dependence on functionality. However, higher values of $M_c/M_c^o$ still result for trifunctional systems at similar values of $M_c^o$ or $\nu$.

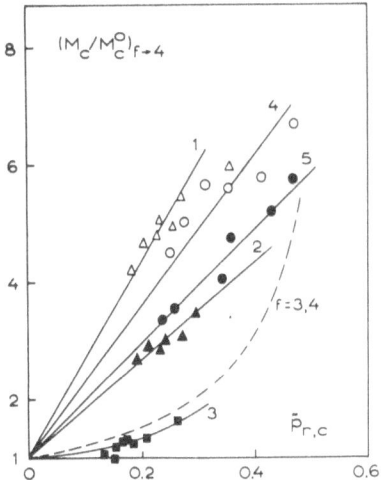

Figure 4.  Normalized molar mass $((M_c/M_c^o)_{f \to 4})$ between elastically
effective junction points versus fractional extent of
intramolecular reaction at gelation $(\widetilde{p}_{r,c})$.  Legend as
for Figure 2 with a single calculated curve for trifunctional and tetrafunctional systems.


The plots in Figure 3 are an attempt to compare the effects of
loops on the moduli of networks of different functionalities on the
approximate bases that the decrease in modulus is in proportion to
the number of smallest loops produced $((M_cM/_c^o)_{3 \to 4})$ and that this
number is proportional to the fraction of pre-gel intramolecular
reaction that has occurred on the basis of the total possible $(\widetilde{p}_{r,c})$.
The correlation sought was that the rescaled $M_c/M_c^o$ for similar systems should depend solely on $M_c^o$ and be independent of f.  This correlation is not achieved and obviously a more detailed consideration
of the types of ring structures introduced by intramolecular reaction
is needed.  However, Figures 2 and 4 do show that the reduction in
modulus is closely related to the amount of pre-gel intramolecular
reaction, with perfect networks being formed only in the limit of
the ideal gel point.  In addition, the reduction in modulus is dependent on the chain structure of the network; compare system 3 with
1, 4 and 5, all of which have similar values of $\nu$.  There is also a
dependence on functionality (system 1 cf. systems 4 and 5) and for
systems of the same functionality and similar chain structure, the
reduction is dependent on the size of the loop that can be formed,
that is, on $M_c^o$ or $\nu$.


Glass Transition Temperature.  The variation of Tg with $\alpha_c$ was
investigated[18] for dry networks formed from systems 3 of Figure 2
at different initial dilutions of reaction mixtures.  Measurements
were carried out at 1 Hz using a torsion pendulum.  The results are

shown in Figure 5. The two limiting values of Tg for this system correspond to networks with $M_c = M_c^O$ and $M_c = \infty$. Thus, the horizontal broken line gives the minimum Tg, that of a linear MDI/POP polymer having a repeat unit of molar mass equal to $M_c^O$, and the maximum value of Tg at $\alpha_c = 0.5$ was obtained by extrapolation of $(1/M_c, Tg)$ data to $1/M_c^O$. The variation of Tg with $\alpha_c$ (or $M_c$) is a reflection of the influence of junction-point density on the freedom of segmental motion. The maximum range of Tg values shown, 301 to 312K (28 to 39°C), possibly reflects the maximum influence for these MDI/POP triol systems. (Following a more detailed analysis of the results,[18] the range of values of Tg is slightly different to that previously quoted.[17])

Properties of Bulk Systems. The results in Figures 2 and 5 clearly show the dependence of materials properties on reaction conditions (initial dilution), on molar mass and functionalities of reactants and on their chain structures. In the context of the commercial preparation of polyurethanes, catalyzed systems reacting in bulk would probably be used. However, provided catalysts and the reaction exotherm result in negligible side reactions, the results are independent of the rate at which a reaction is carried out. Importantly, even in bulk systems, the effects of intramolecular reaction and reactant characteristics are still apparent.

Table I gives shear moduli and Tg's of the networks prepared by bulk reaction of the systems in Figure 2. The first five columns define the systems, the next two give the experimental values of G (at 298K) and Tg, and the last three give the values of $\tilde{p}_{r,c}$, $M_c$, and $G/G^O$ the reduction in rubbery shear modulus on the basis of that

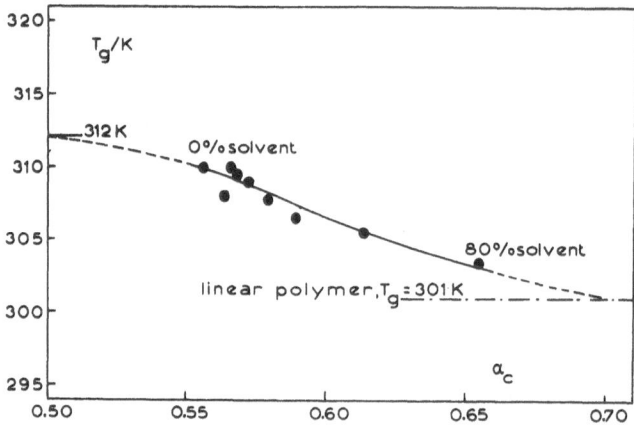

Figure 5.  Tg versus $\alpha_c$ for dry, trifunctional PU networks (●).
Reaction systems:  system 3 of Figure 2 – MDI/LHT240,
$M_c^O$ = 710 g mol$^{-1}$, $\nu$ = 30, prepared at various initial
dilutions of reactants.
–·–·–·– MDI/POP diol, M repeat = $M_c^O$.

Table I

Modulus and Tg of PU Networks from Bulk Reactions

Explanation of systems for Figure 2.  Gel points characterized by $p_{r,c}$. G(298K) is the shear modulus at 298K;  (a) from uniaxial compression of rubbery networks, (b) from torsion pendulum measurements at 1 Hz.  G/G$^o$ is the rubbery shear modulus relative to that expected for the perfect network.

| System | f | $(\alpha_c^o)^{\frac{1}{2}}$ | $\nu$ | $M_c^o$/g mol$^{-1}$ | G(298K)/Nm$^{-2}$ | $T_g$/K | $p_{r,c}$ | $M_c$/g mol$^{-1}$ | G/G$^o$ |
|---|---|---|---|---|---|---|---|---|---|
| 1. HDI/LHT240 | 3 | 0.71 | 33 | 635 | $1.0 \times 10^6$(a) | 255 | 0.183 | 3024 | 0.21 |
| 2. HDI/LHT112 | 3 | 0.71 | 61 | 1168 | $0.8 \times 10^6$(a) | 228 | 0.190 | 3650 | 0.32 |
| 3. MDI/LHT240 | 3 | 0.71 | 30 | 705 | $3.0 \times 10^9$(b) | 311 | 0.117 | 881 | 0.80 |
| 4. HDI/OPPE-NH1 | 4 | 0.58 | 29 | 500 | $1.1 \times 10^6$(a) | 275 | 0.256 | 2273 | 0.22 |
| 5. HDI/OPPE-NH2 | 4 | 0.58 | 33 | 586 | $1.2 \times 10^6$(a) | 271 | 0.238 | 1953 | 0.30 |

expected for the perfect network ($G^O$); $G/G^O$ is in fact equal to $M_c^O/M_c$.

The network from system 3 is distinct from the rest, being a glass at room temperature and also having a rubbery shear modulus near the value expected on the basis of $G^O$. Possible reasons for this high value of $G/G^O$ follow those discussed previously with reference to $M_c/M_c^O$ and Figure 2. The more flexible chains of the aliphatic systems give lower values of Tg, resulting in elastomers at room temperature. Tg is higher for tetrafunctional networks than for the trifunctional network with a similar value of $M_c^O$. In addition, for a given functionality Tg decreases as $M_c^O$ increases. However, remembering Figure 5, Tg will also decrease as $M_c$ (or $\alpha_c$) increases. The latter effect may be of secondary importance as even the smallest ring structures in the present systems are too large to be considered as side groups, and Tg remains related principally to chain structure and the length between <u>chemical</u> junction points. Thus for systems 4 and 5, $M_c^O$ and $M_c$ go in opposite directions, and Tg decreases as $M_c^O$ increases.

The shear moduli of systems 1 and 2 decrease as $M_c^O$ increases, but because of the larger effects of loops in system 1, G for that system is relatively smaller than would be expected on the basis of $M_c^O$. The larger effects are reflected in the higher slopes for system 1 in Figures 2 and 4, as compared with system 2. Thus, G for system 2 is 20% smaller than for system 1, $M_c$ is 20% larger, yet $M_c^O$ is 84% larger. Also the greater, relative decrease in modulus for system 1 occurs in spite of the higher fraction of pre-gel intramolecular reaction ($\widetilde{p}_{r,c}$) in system 2. In addition, with regard to $\widetilde{p}_{r,c}$, because system 2 has a higher value of $M_c^O$ or $\nu$ than system 1, it might be expected that less intramolecular reaction would occur in system 2. However, as discussed in the following section, $\widetilde{p}_{r,c}$ will increase with decreases in both $\nu$ and the concentration of reactive groups. The higher molar masses of the reactants of system 2 mean that in bulk this system is a more dilute system than system 1 with respect to reactive groups, and, hence, more pre-gel intramolecular reaction occurs.

For systems 4 and 5, $\nu$ and $M_c^O$ are closer to each other than for systems 1 and 2. Hence, the concentrations of reactive groups in bulk are more nearly equal and the trend in $\widetilde{p}_{r,c}$ follows that in $\nu$. This produces more rings in system 4, which increases $M_c$ and reduces G relative to system 5. Additionally, the ring structures in system 4 (of lower $M_c^O$) yield a higher proportion of elastically ineffective loops, as indicated by the higher slopes in Figures 2 and 4 for system 4 compared with system 5.

<u>General Comments.</u> The physical properties of network materials formed by random polymerizations, as exemplified by shear modulus and Tg, are markedly affected by the network defects introduced by

the pre-gel intramolecular reaction which has occurred during their formation. The sensitivity of properties to pre-gel intramolecular reaction is higher for networks formed from reactants of lower molar mass ($M_c^o$) and lower functionality, and more flexible chain structures. The physical properties of dry networks can be varied systematically by carrying out preparations at different dilutions of reactants, thus varying $\widetilde{p}_{r,c}$. In this way, the range of properties of materials which can be prepared from given reactants can be extended. This approach to varying properties is possibly not attractive in terms of RIM as it requires the removal of solvent (and some sol fraction). However, it may offer advantages in preparations by casting in moulds.

For bulk reaction mixtures, the modulus of networks with relatively flexible chain structures can be reduced by a factor of five below that expected for network formation in the absence of pre-gel intramolecular reaction. The factor by which G is reduced depends on $M_c^o$, f, chain stiffness, and the initial concentrations of reactive groups obtainable in bulk, in a manner which still needs to be resolved in detail. For systems of stiffer chain structure, there is a lower reduction in modulus than that predicted on the basis of $M_c^o$.

## (ii) Gelation and Pre-gel Intramolecular Reaction Studies

The factors affecting pre-gel intramolecular reaction have been investigated in two ways. First, by the determination of the number of ring structures per molecule as a function of extent of reaction for HDI/POP triol reaction systems at different initial dilutions.[6,27] Second, gelation studies have been carried out using a variety of polyurethane- and polyester-forming reactions,[5-7,10-21,27] with reactants chosen to have like functional groups of equal reactivity. The more detailed results obtained from the first type of investigation have enabled theories of random polymerization which include intramolecular reaction to be investigated.[9,10,17] Eventually, the interpretation of such results is essential if the properties of the network materials formed at complete reaction are to be predicted from reaction conditions (concentration and temperature) and reactant structures. However, for the present purpose of describing the factors which affect $\widetilde{p}_{r,c}$, we need only consider in detail the second type of investigation which focuses solely on the gel point.

The factors influencing intramolecular reaction are the ratio of reactants, the functionalities and the ring forming parameter $\lambda$, where[9,10,13,17,19]

$$\lambda = Pab/c_o, \tag{7}$$

with $c_o$ the initial concentration of reactive groups of one of the reactants and

$$Pab = (3/2 \ \pi\nu b^2)^{3/2}/N \ (\text{moles functional groups/unit volume}, \quad (8)$$

N is Avogardo's number and Pab is the probability that the groups at the two ends of a chain are close enough to react. Pab expresses the dependence of intramolecular reaction on $\nu$, the number of bonds in the chain giving the smallest ring structure that can form, as defined in Figure 1(c), and b is the effective bond length of the chain of $\nu$ bonds, defined such that $\langle r^2 \rangle = \nu b^2$, where $\langle r^2 \rangle$ is the mean-square end-to-end distance of the chain. The approximations involved in the use of Pab have been discussed previously.[13,19] They rest principally on the assumption of Gaussian statistics for the chain of $\nu$ bonds, and its homologues of $2\nu$, $3\nu$, $\cdots\cdots$ bonds, whether they occur in a branched molecule or not, and the use of a constant value of b, independent of chain length.

The allowance for intramolecular reaction en route to gelation leads to the following expression for the gelation condition of $RA_2 + RB_f$ polymerizations[19]

$$\alpha_c(f - 1)(1 - \lambda_{ab})^2 = 1 \tag{9}$$

The ring forming parameter $\lambda_{ab}$ is given by the equation

$$\lambda_{ab} = \frac{(f - 2)Pab.\phi(1,3/2)}{(f - 2)Pab.\phi(1,3/2) + (c_a' + c_b')} \tag{10}$$

$\phi(1,3/2)(= \sum\limits_{i=1}^{\infty} 1^{i} i^{-3/2} = 2.612)$ accounts for the possibility of forming structures from chains of size $\nu$, $2\nu$, $3\nu$, $\cdots\cdots$ bonds, and $(c_a' + c_b')$ is the concentration of reactive groups external to a given molecule, corresponding to $c_o$ in equation (7), giving the probability of intermolecular reaction. $(c_a' + c_b')$ varies as a polymerization proceeds, but in the context of approximate theories of gelation,[13,19] leading to analytical equations such as equation (9), a single, average value of $(c_a' + c_b')$ must be taken. $(c_a' + c_b')$ may most simply be equated to $(c_{ao} + c_{bo})$ and $(c_{ac} + c_{bc})$, the initial and gel-point concentrations of reactive groups, respectively, to give extreme estimates of its value. Equation (9) was developed from the earlier Frisch[28] and Kilb[29] theories of gelation, following a reinterpretation[13] of those theories, according to which equation (10) may be viewed in general terms as

$$\lambda_{ab} = c_{int}/(c_{int} + c_{ext}) \tag{11}$$

$c_{int}$ is the concentration of B groups which can react to form ring structures with a given A group from the same molecule, or vice versa, and $c_{ext}$ is the concentration of groups from other molecules around a given group, that is, an external concentration.

Unlike earlier expressions,[13,28,29] equation (9) does not assume $c_{int} \ll c_{ext}$, so that equations (10) and (11) can be rearranged to give the parameter[19]

$$\lambda'_{ab} = \lambda_{ab}/(1 - \lambda_{ab}) = c_{int}/c_{ext},\tag{12a}$$

where

$$c_{int}/c_{ext} = (f - 2)Pab.\phi(1,3/2)/(c'_a + c'_b)\tag{12b}$$

Figures 6 and 7 show gelation results for the systems discussed in the previous section, with $\lambda'_{ab}$ plotted versus the initial and gel dilutions of reactive groups, respectively. According to equations (12a) and (12b) the plots should be linear with slopes equal to $c_{int}$. Deviations from linearity occur, with the plots for tetrafunctional systems (4 and 5) being the more curved. Corresponding plots for $RA_2 + RB_3$ polyester-forming reactions,[19] for which smaller values of $\lambda'_{ab}$ result, do show linear behaviour when $(c_{ao} + c_{bo})^{-1}$ is used as abscissa and only slight curvature when $(c_{ac} + c_{bc})^{-1}$ is used. The inadequacies of the theory obviously become more apparent the larger the values of $\lambda'_{ab}$ (or $p_{r,c}$).

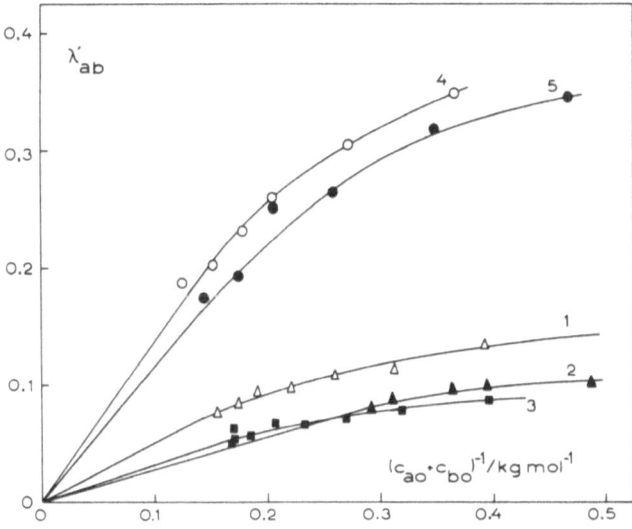

Figure 6.    Ring-formation parameter $(\lambda'_{ab})$ versus dilution of reactive groups $((c_{ao} + c_{bo})^{-1})$. $\lambda'_{ab}$ evaluated from experimental values of $\alpha_c$ according to equations (9) and (12). Abscissa corresponds to $(c'_a + c'_b) = (c_{ao} + c_{bo})$. Reaction systems as for Figure 2. Reactions carried out at 80° bulk and in nitrobenzene solution.

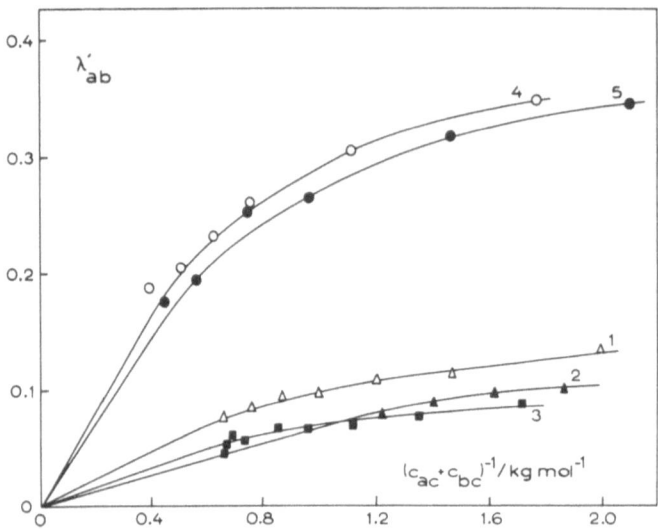

Figure 7.  Ring-forming parameter ($\lambda'_{ab}$) versus gel dilution of
reactive groups (($c_{ac} + c_{bc})^{-1}$).  $\lambda'_{ab}$ values as for
Figure 6.  Abscissa corresponds to ($c'_a + c'_b$) = ($c_{ac} + c_{bc}$).
Reaction systems and reaction conditions as for Figure 6.


In spite of the curved plots, their initial slopes for systems
1, 2, 4 and 5 do decrease as $\nu$ increases, and, if these slopes are
analyzed according to equation (12), the values of b in Table II are
obtained.  In general, b characterizes chain stiffness, and the values in Table II vary with reactant structures and molar masses in
the manner expected and are consistent with values for polyester-[19]
and other polyurethane-forming systems.[17,19,27]  However, the theory
probably undercounts the opportunities for intramolecular reaction
for tetrafunctional systems, relative to trifunctional systems.  Thus,
in compensation, smaller values of b result for tetrafunctional systems in comparison with the values for a trifunctional system with
a similar value of $\nu$ and a similar chain structure (systems 4 and 5
cf. system 1).  However, compared with the values of b obtained using
earlier gel-point expressions,[13,28,29] much better agreement between
values for trifunctional and tetrafunctional systems of similar chain
structures is achieved on the basis of equation (12).

In Table II, $\nu_{DI}/\nu$ is the fraction of bonds due to the diisocyanate residue in the chain of $\nu$ bonds.  This residue will be stiffer (have larger value of b) than that due to the oxypropylene units.
In fact, with regard to the HDI-based systems $(<r^2_{o,n}>/n)^{1/2}$ = 0.40 nm
for an infinite, linear polymethylene (PM) chain and 0.34 nm for an
infinite, linear POP chain.[30]  Larger values of b than for PM would
be expected for the hypothetical linear chain based on MDI.  Thus,

Table II

Values of Effective Bond Length (b) of Chains Forming the Smallest Ring Structures (of $\nu$ bonds), Derived from Figures 6 and 7 and Equation (12)

(i) $c_{ext} = (c_{ao} + c_{bo})$, (ii) $c_{ext} = (c_{ac} + c_{bc})$

$\nu_{DI}/\nu$ is the fraction of bonds due to the diisocyanate residue in a chain of $\nu$ bonds.

| | System | f | $(\alpha^o_c)^{\frac{1}{2}}$ | $\nu$ | $\nu_{DI}/\nu$ | b/ nm (i) | b/ nm (ii) |
|---|---|---|---|---|---|---|---|
| 1. | HDI/LHT240 | 3 | 0.71 | 33 | 0.303 | 0.247 | 0.400 |
| 2. | HDI/LHT112 | 3 | 0.71 | 61 | 0.164 | 0.222 | 0.363 |
| 3. | MDI/LHT240 | 3 | 0.71 | 30 | 0.233 | 0.307 | 0.488 |
| 4. | HDI/OPPE-NH1 | 4 | 0.58 | 29 | 0.345 | 0.240 | 0.356 |
| 5. | HDI/OPPE-NH2 | 4 | 0.58 | 33 | 0.303 | 0.237 | 0.347 |

for systems 1 and 2, b decreases as $\nu_{DI}/\nu$ decreases, and the same
trend is apparent for systems 4 and 5. (For these pairs of systems,
the same trend is in fact apparent on the basis of $\nu$ alone. However, results for systems encompassing different aliphatic difunctional components[5,12,13,19] have shown that it is the fractional
chain length due to the difunctional component $(\nu_2/\nu)$, rather than
$\nu$, which generally correlates with b.) Finally, as expected, the
MDI-based system has larger values of b than the trifunctional HDI-
based systems.

In general, the plots in Figures 6 and 7 show that the major
factors affecting intramolecular reaction are the dilution of a reaction system and the value of $\nu$, while the values of b in Table II
illustrate a secondary dependence on chain structure. The absolute
values of b given in Table II are extreme values, resulting from the
extreme values of $c_{ext}$ employed. The true values of b should lie
somewhere between the two and probably nearer the values obtained
using $c_{ext} = (c_{ac} + c_{bc})$, as the opportunities for intramolecular
reaction increase as the gel point is approached. Recalling the
values of b of 0.40 nm and 0.34 nm for PM and POP chains, the values
of n in the last column in Table II for the trifunctional, HDI-based
systems are in keeping with this expectation, whereas that for the
trifunctional MDI-based system is correspondingly higher, and those
for the tetrafunctional, HDI-based systems somewhat lower than expected.

It should be remembered that the experimental points at the
lowest dilutions in Figures 6 and 7 refer to bulk reaction systems,
and the preceding interpretations of initial slopes in terms of $\nu$
and b is in keeping with the gelation behaviour of such systems.
Thus, in spite of the deviations from the predictions of equations
(9) and (12) at higher dilutions, these equations can be used to
correlate gel points of bulk or concentrated systems in terms of
reactant structures. In view of the correlations established in
the preceding section between gel point and the properties of the
materials formed at complete reaction and the correlations demonstrated in the present section between gel point and reactant structures, direct correlations between reactant structures and materials
properties can be established, at least for bulk systems. Such correlations are relevant to RIM, where the gel point, although an important parameter in terms of in-mould flow, should be purely an
incidental point in the process. Thus, in the context of RIM, the
present work has illustrated the possibility of correlations between
reactant structures and materials properties, for model systems, with
like groups having equal reactivities and with equimolar mixtures
of reactive groups. That intramolecular reaction cannot usually be
neglected for non-linear polymerizations in bulk has been established
by gelation studies on a wide range of systems,[5-7,10-21,27] and from
the determination of the number of ring structures per molecule as
a function of pre-gel intramolecular reaction.[6,27] The behaviour

of non-linear bulk reaction mixtures is in marked contrast with that of bulk linear systems.[6] For example, for a bulk HDI/POP triol system with $\nu = 115$ about one molecule in three had a ring structure at the gel point,[6] whereas a bulk, HDI/POE diol system with $\nu = 25$ gave negligible ring fractions even at high extents of reaction.[4]

### (iii) Correlations Between Reactant Characteristics and Materials Properties

Sections (i) and (ii) have considered model PU network forming systems in which, because of the relatively slow rates of reaction, it has been possible to determine the amount of pre-gel intramolecular reaction and study its influence on materials properties. With regard to direct applications to RIM, the reactions systems and networks formed may be unsuitable for commercial exploitation because of the slow rates of reaction and the physical properties obtainable. However, slight modifications of the systems can make them suitable for use. This may be illustrated by the results of studies on unfilled and filled PU materials based on system 3 with the inclusion of trimethylol propane (TMP). TMP, which contains primary hydroxyl groups, thereby reduces the gel time. Table III gives the gel times for bulk reactions for systems 1 to 5 and for system 6, namely, system 3 with a particular LHT240:TMP molar ratio. In can be seen that the addition of catalyst to system 6 could reduce the gel time to the order of those of current commercial interest in RIM (15 to 30s).

Several variants of system 6 using different LHT240:TMP molar ratios have been studied. The network materials have all been prepared by reactions at 50°C in bulk, and with stoichiometric amounts of -NCO and -OH groups ($r = 1$). The Tg network materials formed at complete reaction varied from 311K (38°C) to 453K (180°C) as the LHT240:TMP ratio varied from 100:0 (system 3) to 5:95. Some LHT240 was essential to keep the reactants miscible, and in all cases, clear, single-phase materials were formed. For those systems with Tg's in excess of the reaction temperature, post-curing at above Tg was necessary to take reactions to completion.

The increase in Tg with increasing proportion of TMP is to be expected because TMP is of lower molar mass than LHT240 and results in a higher density of junction points. Thus, the trend is in keeping with that in Figure 5 where Tg increases as $\alpha_c$ (or $M_c$) decreases. The reason for the variation of Tg with $M_c$ in Figure 5 was network defects; the reason for the present variation is the systematic change in $M_c^O$ with proportion of TMP. Effects of intramolecular reaction, which will lower the absolute Tg's below those expected for perfect networks, are neglected. In fact, more intramolecular reaction would be expected for a higher proportion of TMP ($M_c^O$ and $\nu$ are reduced) causing a larger reduction in Tg compared with the value expected for a perfect network. However, the predictions of

Table III

Gel Times for Bulk Reactions of Systems 1 to 6

All reactions at r = $NCO_o/OH_o$ = 1

Systems 1 to 5 as Figure 2. System 6, based on system 3 with the addition
of trimethylol propane (TMP) to give a molar ratio of LHT240:TMP of 10:90

| System | ( $NCO_o$ + $OH_o$ )/ mol $kg^{-1}$ | Reaction temperature/ $^o$C | Gel time/ min |
|---|---|---|---|
| 1. HDI/LHT240 | 6.2 | 80 | 830 |
| 2. HDI/LHT112 | 3.4 | 80 | 1600 |
| 3. MDI/LHT240 | 5.6 | 80 | 45 |
| 4. HDI/OPPE–NH1 | 8.0 | 80 | 224 |
| 5. HDI/OPPE–NH2 | 6.8 | 80 | 323 |
| 6. MDI/LHT240(10)/TMP(90) | 10.6 | 50 | 5 (vitrification) |

ideal gel points and the expected properties of perfect networks are
more difficult with the present formulations as they contain hydroxyl
groups of different reactivities.  From Figure 5 the range of Tg's
obtainable by varying the amount of intramolecular reaction (303 to
311K) is seen to be an order of magnitude less than that obtainable
from the present systems (311 to 453K) by varying $M_c^o$.  Thus, it is
not surprising that the overall trend in Tg follows that in $M_c^o$.

At ambient temperatures, tough, stiff, glassy materials result
from the use of system 6 and variations thereof, with typical Young's
moduli of 3 GN m$^{-2}$ and tensile strengths approaching 80 MN m$^{-2}$.  In
comparison with the high-modulus RIM polyurethanes normally encountered commercially, the materials formed are not only stiffer and
stronger, but are also more thermally and dimensionally stable at
temperatures in excess of 100°C.  The better high-temperature properties result from the permanent network structure which also results in Tg's up to 453K (180°C).  Tough materials are produced despite the high junction-point densities, and the toughness is largely
attributable to the presence of the flexible POP chains in the network structures.  Dynamic-mechanical tests at 1 Hz in the temperature range 113K (-160°C) to 453K (180°C) showed for system 6 the
presence of three, merged secondary relaxations at 183K (-90°C),
263K (-10°C) and 338K (65°C), giving a broad damping region from 173
to 373K (-100 to + 100°C).  These damping characteristics are largely responsible for the good fracture toughness observed for this
material under impact loading.  Using notched-Charpy specimens, impact strengths of the order 6.1 kJ m$^{-2}$ were observed.  (cf. 1.1
kJ m$^{-1}$ measured for commercial poly(methylmethacrylate) using identical tests.)

In the context of RRIM, materials formed from system 6 in the
presence of hammer-milled glass (mean length 70 μ m) and chopped-
strand (nominal length 1.5 mm) have been evaluated.  Figure 8 shows
the expected increase in modulus with weight fraction of filler.
For a weight fraction of 30% of hammer-milled glass, the modulus is
increased from 2.8 to 5.9 GN m$^{-2}$ with a concurrent increase in tensile strength from 77 to 92 MN m$^{-2}$.  On the basis of the notched-
Charpy test, the expected decreases in impact strength with filler
loading are observed; 6.1 to 2.5 kJ m$^{-2}$ at 30% w/w hammer-milled
glass.  Nevertheless, the filled material still exhibited a high degree of fracture toughness as indicated by the fracture surface,
produced after impact failure, and shown in Figure 9.  The micrograph shows a very rough fracture surface in which the fibres have
been well dispersed and fully wetted by the PU matrix.  Despite the
short length of the hammer-milled fibres, there is evidence of some
broken fibre-ends indicating a strong degree of bonding developed
between the fibres and the PU matrix.  With chopped-strand glass,
problems due to high viscosity and dispersion limit the loading of
fibres which can be achieved in these materials.  Nevertheless, at
equivalent loadings to the hammer-milled glass, greater increases

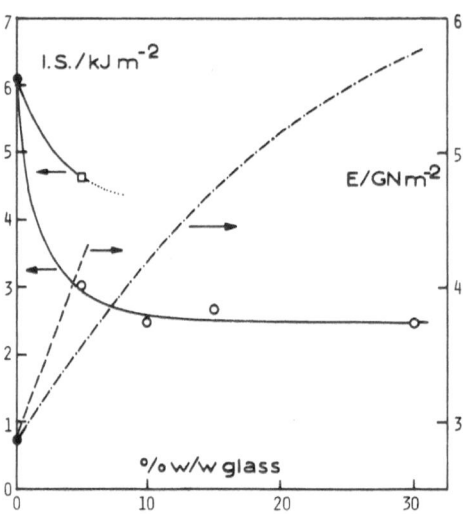

**Figure 8.** Impact strength (I.S.) and Young's modulus (E) versus
filler content for composite materials based on system
6 (MDI/LHT240(10)/TMP(90). Unfilled matrix: o
Hammer-milled glass fibre: I.S., O; E, –•–•–•–
Chopped-strand glass fibre: I.S., : E, –––––––

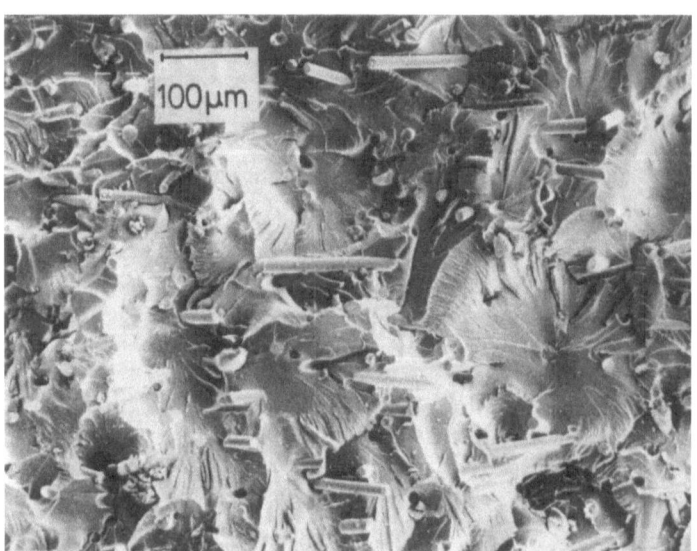

**Figure 9.** Micrograph of fracture surface resulting from impact
failure. Space PU-composite formed from system 6 with
hammer-milled glass.
Tg = 423K (150°C). I.S. = 4.6 kJ m$^{-2}$.

in modulus (and strength) and smaller decreases in impact strength were observed, as shown in Figure 8.


SUMMARY

    A summary of polyurethane (PU) network materials including those formed by fast-reacting, model systems suitable for use in RIM have been presented. The reaction systems comprise commercially available diols and triols and especially synthesized tetrols, and aliphatic and aromatic diisocyanates.

    By varying polyol functionality, and the chain structures and concentrations of reactants, it has been shown that the physical properties of the materials formed at complete reaction may be correlated in terms of these parameters. The product of the extents of reaction at gelation has been demonstrated to be a useful intermediate parameter in this correlation.

    With reference to reinforced reaction injection moulding (RRIM), studies of composite materials formed from PU networks and inorganic fillers have been presented, and their properties related to those of the PU matrices and fillers.


References

1.  "Reaction Injection Moulding," Van Nostrand Reinhold Co., 1979, ed. W. E. Becker.
2.  E. Broyer and C. W. Macosko, Amer. Inst. Chem. Eng., J. 1976, 22, 268.
3.  S. D. Lipshitz and C. W. Macosko, Polym. Eng. & Sci., 1976, 16, 803.
4.  R. F. T. Stepto and D. R. Waywell, Makromol. Chem., 1972, 152, 247, 263.
5.  W. Hopkins, R. H. Peters and R. F. T. Stepto, Polymer, 1974, 15, 315.
6.  J. L. Stanford and R. F. T. Stepto, Brit. Polym. J., 1977, 9, 124.
7.  N. G. K. Hunt, R. F. T. Stepto and R. H. Still, Proc. IUPAC Int. Symp. on Macromolecules, Dublin, 1977, p. 723.
8.  J. L. Stanford and R. F. T. Stepto, J. Chem. Soc. Faraday Trans. I., 1975, 71, 1292.
9.  J. L. Stanford, R. F. T. Stepto and D. R. Waywell, J. Chem. Soc., Faraday Trans., 1975, 71, 1308.
10. J. L. Cawse, J. L. Stanford and R. F. T. Stepto, Proc. IUPAC Int. Symp. on Macromolecules, Mainz, 1979, p. 693.
11. R. H. Peters and R. F. T. Stepto, "Chemistry of Polymerization Processes," Soc. Chem. Ind. Monograph No. 20., London, 1965, p. 157.

12. R. S. Smith and R. F. T. Stepto, Makromol. Chem., 1974, <u>175</u>, 2363.

13. R. F. T. Stepto, Faraday Disc., Chem. Soc., 1974, <u>57</u>, 69.

14. A. B. Fasina and R. F. T. Stepto, Proc. IUPAC Int. Symp. on Macromolecules, Dublin, 1977, p. 717; Macromol. Chem. 1981, in press.

15. J. P. Berry, A. C. Bissett and J. L. Stanford, Proc. IUPAC Int. Symp. on Macromolecules, Dublin, 1977, p. 735.

16. N. G. K. Hunt, R. F. T. Stepto and R. H. Still, Proc. IUPAC Int. Symp. on Macromolecules, Mainz, 1979, p. 697.

17. R. F. T. Stepto, Polymer 1979, <u>20</u>, 1324.

18. J. L. Cawse, Ph.D. Thesis, University of Manchester, 1979.

19. Z. Ahmad and R. F. T. Stepto, Colloid and Polymer Sci., 1980, <u>258</u>, 663.

20. Yoji Hirasawa, R. F. T. Stepto and R. H. Still, Proc. IUPAC Int. Symp. on Macromolecules, Strasbourg, 1981, in press.

21. S. V. Rose and J. L. Stanford, unpublished work.

22. D. P. M. Dunn and J. L. Stanford, unpublished work.

23. K. Dusek and W. Prins, Adv. Polymer Sci., 1969, <u>6</u>, 1.

24. P. J. Flory, Proc. Roy. Soc. (London), 1976, <u>A351</u>, 351.

25. J. E. Mark, The IUPAC Disc. Conf. on Macromolecules (Polymer Networks), Karlovy Vary, 1980.

26. P. J. Flory, "Principles of Polymer Chemistry," Cornell Univ. Press, Ithaca, 1953, Chap. IX.

27. Z. Ahmad, Ph.D. Thesis, University of Manchester, 1977.

28. H. L. Frisch, Preprints Amer. Chem. Soc., 128th Meeting, Polymer Division, Minneapolis, 1955.

29. R. W. Kilb, J. Phys. Chem., 1958, <u>62</u>, 969.

30. P. J. Flory, "Statistical Mechanics of Chain Molecules," Interscience Publishers, New York, 1969, p. 40.

# RIM URETHANES STRUCTURE/PROPERTY RELATIONSHIPS FOR LINEAR POLYMERS

Richard J. Zdrahala and Frank E. Critchfield

Silicones and Urethane Intermediates Division
Union Carbide Corporation
South Charleston, W. Va. 25303

SUMMARY

Reaction Injection Molding (RIM) technology represented the major break-through in the polymer processing development of the seventies. This technology, based on the rapid mixing of co-reactive streams and subsequent feeding of the mixture into the polymerization reactor – the mold, – allows for production of large and intricate parts. The relatively low energy input that is required makes this technology especially attractive for today's needs.

Polyurethanes are perfectly suited for RIM. The fast and exothermic reaction typical for urethane formation does not generate any low molecular weight byproducts. And by the selection of different intermediates, the properties of the resultant polymer can be broadly varied.

In order to investigate the structure/property development of linear RIM urethanes, a series of polyether-based linear (thermoplastic) polyurethanes (TPU) varying in the hard segment content between 20 and 80 weight percent was prepared. 4,4'-Diphenyl-methane diisocyanate (MDI) and 1,4-butanediol (BDO) were the hard segment intermediates. (Oxypropylene-oxyethylene) diols of $\overline{M}_n$ = 1000, 2000, 3000 and 4000 were used as the soft segments. Physical-mechanical, dynamic-mechanical and specific heat (DSC) are used to elucidate the mechanical and morphological behavior of the materials. The polyurethanes varied from soft elastomerics (continuous soft phase) to high-modulus plastics (continuous hard phase) with a possible phase inversion at about 60 percent hard segment.

EXPERIMENTAL

The linear polyurethanes were prepared by a hand-casting procedure described in detail elsewhere.[1]  The series of four polyether
polyols having $\overline{M}_n$ of 1000, 2000, 3000 and 4000 were used as the soft
segments.  All of these polyols contained 15 weight percent of oxy-
ethylene as end cap on an oxypropylene backbone.  The polyols were
dried over molecular sieves (Linde, Type 4A) for about a week prior
to use.  Anhydrous 1,4-butanediol was vacuum distilled (5 torr/95°C)
and stored over molecular sieves.  The MDI was freshly filtered
(~50°C) prior to use and only water-white material was used.  The
polymers were post-cured for 16 hours at 100°C and allowed to "age"
for one week at ambient temperature prior to testing.

Standard physical-mechanical properties were measured at ambient
conditions with an INSTRON Tensile Tester in accordance with prescribed ASTM methods (D638-72).

The differential scanning calorimeter consisted of a duPont 990
Thermal Analyzer with a Model 910 cell base.  A PDP 11/40 minicom-
puter was used to collect the analogue signals ($T_1 \frac{dq}{dt}$) in real time,
perform the required data reduction and plot the resulting curves.[2]
DSC measurements were made at 10°C/min in dry nitrogen environment
using sample weights of approximately 20 mg.

A Rheovibron, Model 200, manufactured by Toyo-Baldwin Co., Ltd.,
was used for dynamic-mechanical properties.  Data were obtained at
a frequency of 11 Hz while heating at 2°C/min in dry nitrogen.  Cor-
rections were made for instrument compliance and end effects using
the method of Massa.[3]


RESULTS AND DISCUSSION

Physical-Mechanical Properties

Increasing the molecular weight of the soft segment should promote phase separation and amplify the block character of linear polyurethanes.  This should have a positive influence on some key physical-mechanical properties such as tensile strength, elongation, flexural modulus/temperature relationship, etc.  Indeed, these properties
are significantly influenced as can be seen in Figures 1, 2, 3 and 4.

Figure 1 illustrates the dependence of the tensile modulus at
100% elongation on the soft segment molecular weight.  Essentially,
regardless of the hard segment content, the tensile modulus increases
with an increase in the soft segment molecular weight.  This trend
reverses for polymers having greater than 70 weight percent hard segment, suggesting an inversion of phases.

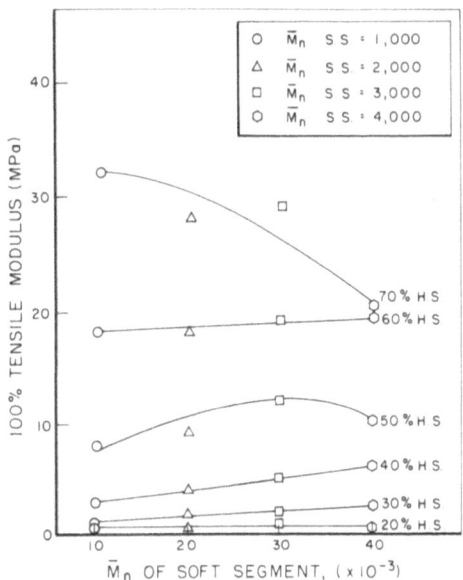

Figure 1.  Dependence of 100% tensile modulus on the soft segment
molecular weight and the hard segment content of
thermoplastic polyurethanes.

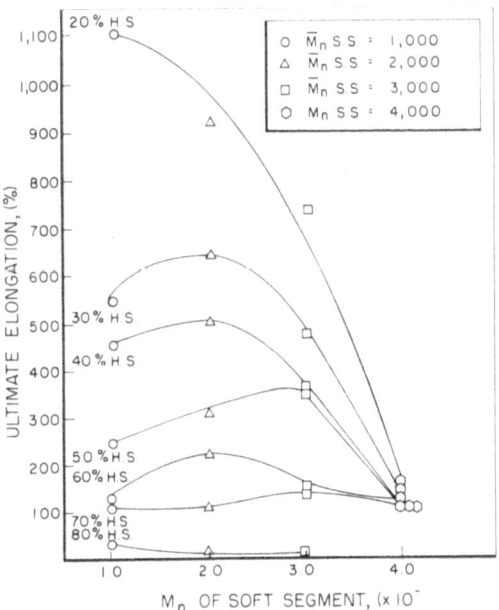

Figure 2.  Dependence of ultimate elongation on the soft segment
molecular weight and the hard segment content of
thermoplastic polyurethanes.

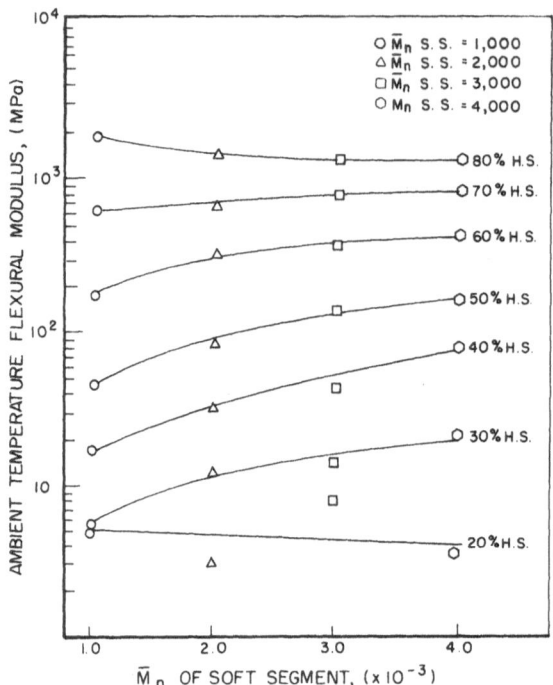

Figure 3.   Dependence of ambient temperature flexural modulus on
            the soft segment molecular weight and hard segment
            content of thermoplastic polyurethanes.


     An interesting relationship was observed for the ultimate elong-
ation as a function of the soft segment molecular weight.  For a particular level of the hard segment, the ultimate elongation seems to
peak when the $\overline{M}_n$ of the soft segment is between 2000 and 3000.  Figure 2 illustrates this trend.

     Among the physical mechanical properties of linear urethanes,
the flexural modulus and the temperature dependence is probably most
closely tied to the structure and morphology of the system.  Since
covalent crosslinks are essentially absent, the "virtual crosslinks"
created by the hard segment domains provide most of the reinforcement.  Thus, increasing the molecular weight of the soft segment,
through its positive effect on phase separation and domain formation,[4] should increase the values of the flexural modulus at a particular level of the hard segment.  This trend reverses at about 70
weight percent of the hard segment, again, suggesting phase inversion.  The results are illustrated in Figures 3 and 4.

Dynamic-Mechanical and Thermal Properties

     In Figure 5, temperature dependence of the tensile storage

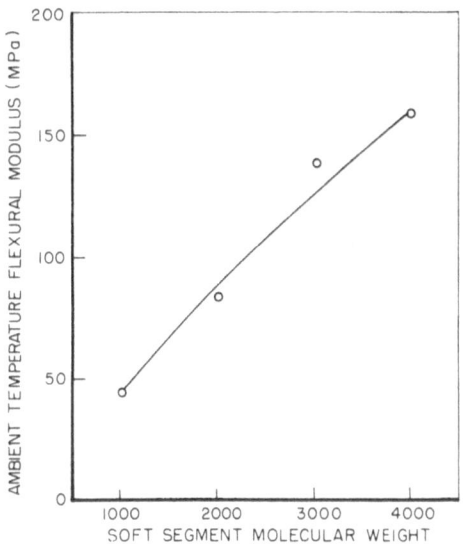

Figure 4.  Dependence of the ambient flexural modulus of thermoplastic polyurethanes on the soft segment molecular
weight at 50% hard segment.

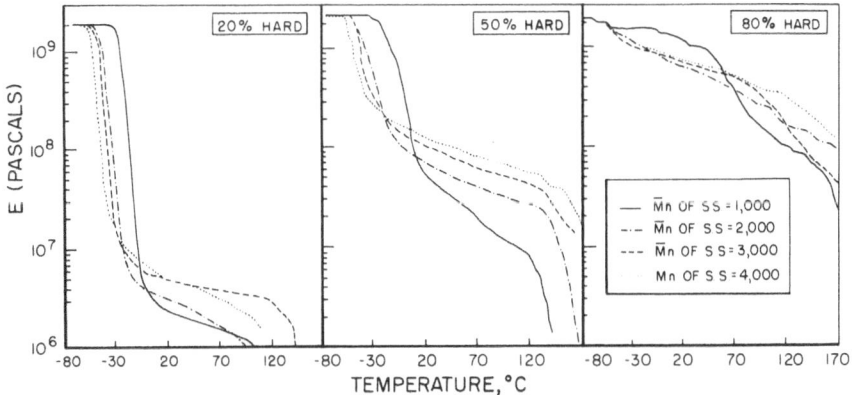

Figure 5.  Elastic modulus vs. temperature curves for TPU's with
1000 (——), 2000 (—·—), 3000 (---) and 4000 (•••)
molecular weight soft segments and 20, 50 and 80%
hard segment content.

moduli (E') are given for the polymers containing 20, 50 and 80
weight percent of the hard segment.  The first two sets clearly
confirm the two-phase character of these urethanes.  A sharp decrease in elastic modulus is clearly visible in the soft segment
Tg region.  The temperature range for the soft segment Tg is shifted
down scale as the soft segment molecular weight increases.  A relatively flat plateau region is seen in the E' vs. temperature curves
above the soft phase glass relaxation.  The modulus in the plateau
region increases with increasing hard segment content and also with
increasing $\overline{M}_n$ of the soft segment.  For most polymers the modulus
plateau extends to about 150°C, where the hard phase disruption
begins.  Because of sample softening, we could not continue through
this transition.

The above observed transitions were confirmed by DSC analyses.
Figure 6 provides examples of DSC curves.  A distinct glass transition due to the soft phase is seen between -65 and 10°C for the polymers with hard segment contents of 60 percent or less.  The transition temperature does not change appreciably with hard segment con-

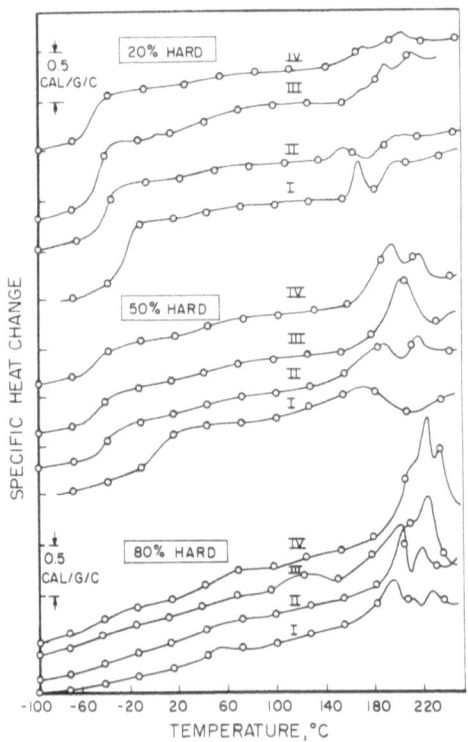

Figure 6.   DSC curves for TPU's with 1000 (I), 2000 (II), 3000 (III)
and 4000 (IV) molecular weight soft segments and 20, 50
and 80% hard segment content.

tent but decreases with increasing soft segment molecular weight.
A second weak inflection is seen in the 20 to 80°C region, probably
due to phase mixing at the domain interface.  Finally, endothermic
peaks are observed between 140 and 230°C for all of the polymers.
There is considerable evidence in the literature[5,6,7] supporting the
assignment of these peaks to melting of crystalline hard segment.
Figures 7 and 8 quantify the shifts in the soft segment Tg's and
hard segment Tm's with respect to the soft segment molecular weight
and the hard segment content.


CONCLUSIONS

    This study illustrates the structure/property relationships of
linear polyurethanes suitable for the RIM technology as a function
of the hard segment content and the soft segment molecular weight.
Firstly, increasing the hard segment content increases the strength
and modulus of the polymer and changes the character of the urethane
from elastomeric to plastic.  And secondly, increasing the soft segment molecular weight also increases the modulus, decreases soft
segment Tg and increases Tm of the urethane.

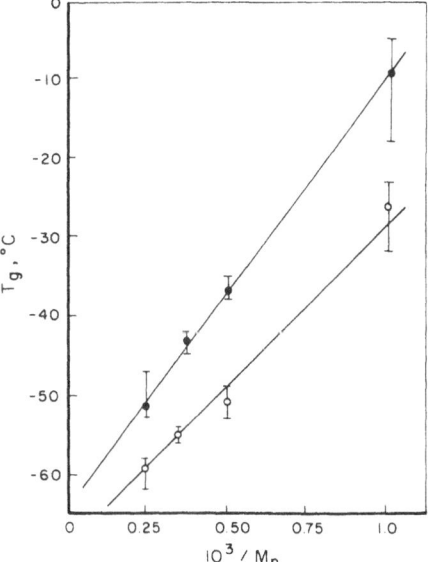

Figure 7.  Dependence of soft phase Tg on soft segment molecular
           weight.  20-60% hard segment content.
           0 = average Tg from DSC.
           0 = average Tg from E'' (Rheovibron).
           Bars show variation with hard segment content.

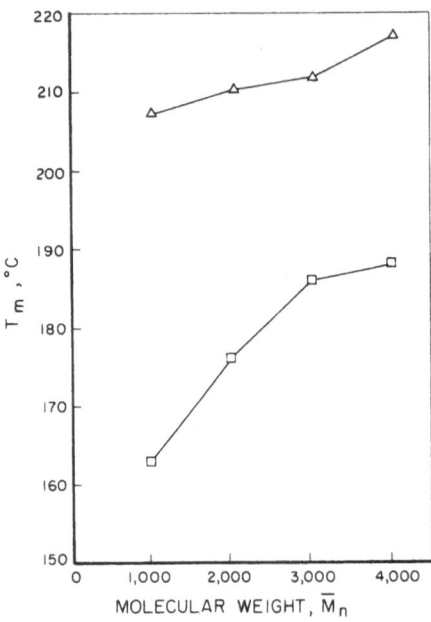

Figure 8.  Dependence of hard phase melting point on soft segment
molecular weight at 30 ($\square$), and 80% ($\Delta$) hard segment
content.

## References

1. R. J. Zdrahala, R. M. Gerkin, S. L. Hager and F. E. Critchfield,
   J. Appl. Polym. Sci., 24, 2041 (1979).
2. S. L. Hager, Thermochimica Acta, 26, 149 (1978).
3. D. J. Massa, J. Appl. Phys., 44, 2595 (1973).
4. C. G. Seefried, Jr., J. V. Koleske and F. E. Critchfield, J.
   Appl. Polym. Sci., 19, 2503 (1975); ibid, 19, 3185 (1975).
5. L. L. Harrell, Jr., Macromolecules, 2, 607 (1969).
6. T. Kajiyama and W. J. MacKnight, Polymer J., 1, 548 (1970).
7a. A. L. Chang and E. L. Thomas, "Multiphase Polymers," pp. 31-52;
    S. L. Cooper and G. M. Ester, Eds., ACS, Washington, DC, 1979.
7b. C. S. Schollenberger, ibid, pp. 83-96.
7c. J. W. C Van Bogardt, A. Lilaonitkul and S. L. Cooper, ibid,
    pp. 3-30.

THE EFFECT OF HARD SEGMENT CONTENT ON

A CROSS-LINKED POLYURETHANE RIM SYSTEM

R. B. Turner, H. L. Spell and J. A. Vanderhider

Dow Chemical Company

Freeport, TX

INTRODUCTION

Since the development of the Reaction Injection Molding (RIM) process for the manufacture of segmented polyurethane systems and possible use of such polymers in the automobile industry, much emphasis has been placed on the study of property-structure relationships of these polymers.[1-20] The goal of this study is to understand and thus optimize physical property characteristics of such systems containing ethylene glycol as a chain extending agent.

These polyurethane systems are known to be (AB)n multi-block polymers having definite microphase domains ("hard" and "soft").[2,3] These polymers offer several unique properties: (1) minimum physical property loss in temperature ranges from -20°C to 100°C, (2) good toughness, (3) ease of fabrication at rapid cycle times, and (4) RIM processability from available raw materials.

The present study differs from previous works in the literature in three major areas: (1) the polymers are thermosetting rather than thermoplastic polyurethanes, (2) the high molecular weight polyol (soft segment) is a polyether triol rather than the polyester or polyether diols commonly used, and (3) the mode of sample preparation is by commercial RIM equipment. The polymers considered in this work are thought to contain both "hard" and "soft" domain structures and can be considered multi-block copolymers. The hard segments should consist of the reaction product of ethylene glycol (EG) and 4,4'-methylene bis(phenylisocyanate) (MDI). The soft segments should be comprised of a 7200 molecular weight polyether polyol. This polyol is a glycerine initiated triol propoxylated and subsequently end capped with ethylene oxide. The soft segments are attached into the

hard segment systems through MDI coupling of terminal hydroxyl groups
of the polyol with those of the glycol (EG).

Several investigators[3,4] have defined the major contributors to
microphase segregation for these types of polyurethanes. These factors include: (1) extent of hard domain crystallization, (2) steric
hindrance effects on hard domains due to hydrogen bonding, and (3)
solubility differences between the hard and soft domains. The degree
or extent to which the two phases are actually segregated plays a major role in the physical properties of the resultant polymer. In the
present work a triol is used to give a lightly cross-linked soft domain in an effort to enhance phase segregation. The volume percent
hard segment is studied from relatively low levels to a point beyond
phase inversion or the hard and soft domains (the hard domain becomes
the continuous phase). In contrast to early investigations, ethylene
glycol was used instead of 1,4-butanediol as the hard segment forming
glycol.[1] This was done to improve the thermal stability of the final
polymer. Ethylene glycol (EG) produces a hard segment melting point
of about 215-230°C whereas the melting point of the corresponding hard
segment of 1,4-butanediol (BDO) is between 180-195°C. These polymers
were made using RIM process equipment as opposed to hand cast techniques. As a consequence the polymerization process more nearly du-
plicates industrial practice.


EXPERIMENTAL

Polymer Synthesis

The polymers were made by reacting uretoneimine modified 4,4'-
diphenylmethane diisocyanate (MDI) with a polyol blend consisting of
ethylene glycol (EG) and a high molecular weight triol. The polyether triol is initiated with glycerine, reacted with propylene oxide
to 6000 MW and further reacted with ethylene oxide to 7200 MW and has
84% primary hydroxyl end groups (see structures in Table I). These
materials were polymerized using a Krauss-Maffei PU 80/40 RIM (Reaction Injection Molding) machine using a suitable catalyst (Witco Fom-
rez UL-28). 56 cm by 61 cm by 0.32 cm plaques were produced in a P20
steel mold heated to 160°F (71°C).

Dynamic Mechanical Spectroscopy (DMS)

Dynamic mechanical measurements of the six segmented polyurethanes were made with a Rheometrics Mechanical Spectrometer, Model
605. The temperature range was -155°C to 275°C and the oscillatory
frequency was 1 Hz. Strain was held to 0.1% throughout the scan.
After the sample was cooled to approximately -155°C, it was heated
at a constant rate and measurements were taken at 5°C intervals. In
some instances, the intervals were shortened to two minutes at the
high end of the temperature scan. The samples were allowed to soak

Table I

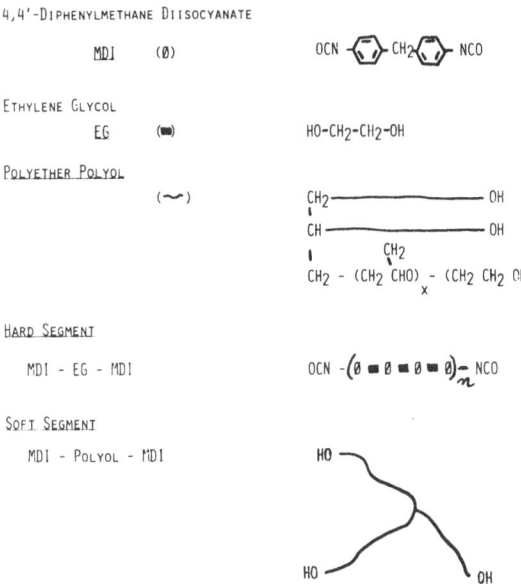

4,4'-Diphenylmethane Diisocyanate

MDI (Ø)

Ethylene Glycol

EG

Polyether Polyol

(〜)

Hard Segment

MDI - EG - MDI

Soft Segment

MDI - Polyol - MDI

for three minutes after the desired temperature was reached before a reading was taken. Dimensions of the sample bars were approximately 62 millimeters by 12 millimeters by 3 millimeters. The polymers were kept in a nitrogen atmosphere throughout the analysis. The measurements were performed with an accuracy of 0.5°C for temperature readings and a precision of 2% for the modulus values.

## Differential Scanning Calorimeter (DSC)

Differential scanning calorimeter (DSC) measurements were taken by the use of a Mettler TA 2000 at a scan rate of 10°C per minute, chart speed of 5 millimeters per minute and an amplifier range of 100 microvolts/full scale deflection. The samples were cooled to -120°C in a nitrogen atmosphere. Those samples that are noted as annealed were placed in an oven at 325°F (163°C) for one hour and the oven was shut off and cooled to room temperature over a period of four hours. The samples were then removed and tested. The sample size was approximately 10 milligrams.

RESULTS AND DISCUSSION

## Dynamic Mechanical Measurements

Dynamic mechanical measurements of the various polymers are shown graphically in Figures 1 and 2. In the temperature range above

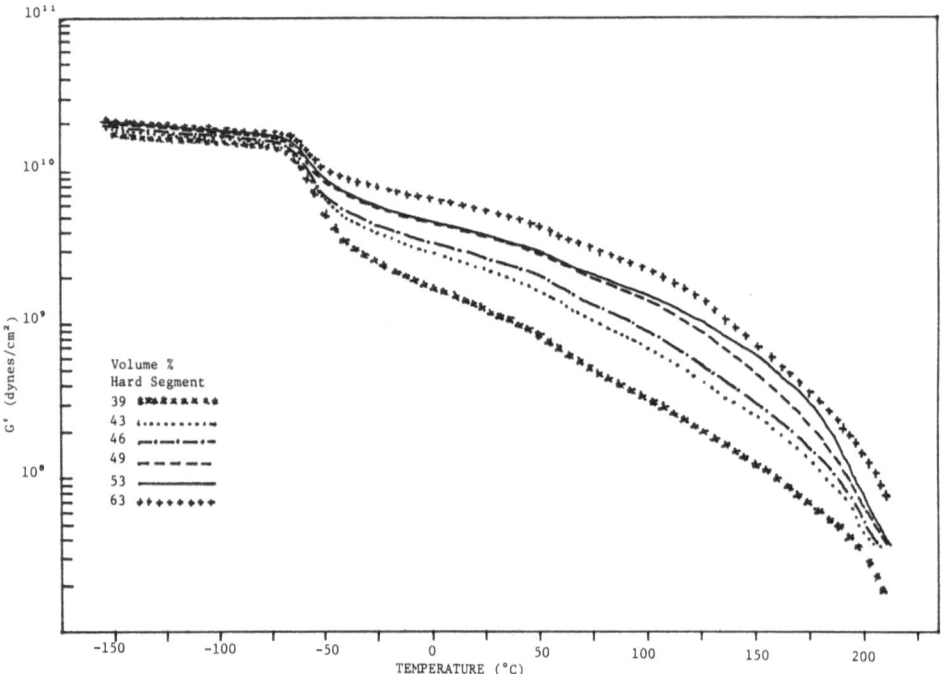

Figure 1.  Effect of hard-segment concentration on the storage
           modulus (G') of segmented polyurethanes.


the glass transition of the soft segment, the storage moduli of all
the samples are about $2 \times 10^{10}$ dynes per square centimeter.  These
values are in agreement with results from other workers.  For the
first five samples which contained from 53 to 39 volume % hard segment, the storage moduli dropped rather rapidly at the glass transition region.  The decreases in modulus at this point correlate
roughly with the decrease in hard segment content.

     The next significant drop in storage modulus occurs at the glass
transition of the hard domain.  From that point (ranging from 30°C to
85°C depending on the hard segment content), the storage modulus
curves are somewhat complex.  Several short decreases occur as the
temperature increases up to the final sharp drop at the melting point
and/or decomposition of the polymers.

     The polymers with very high hard segment content (63 and 73 vol-
ume percent) exhibit storage modulus curves quite different from the
ones previously discussed.  For example, the change in modulus in
going through the soft segment transition is relatively small.  This
as well as the unusually large drop in modulus at about 85°C, may
indicate phase inversion has occurred in these particular materials.
That is to say, they may be characterized as having a fairly contin-

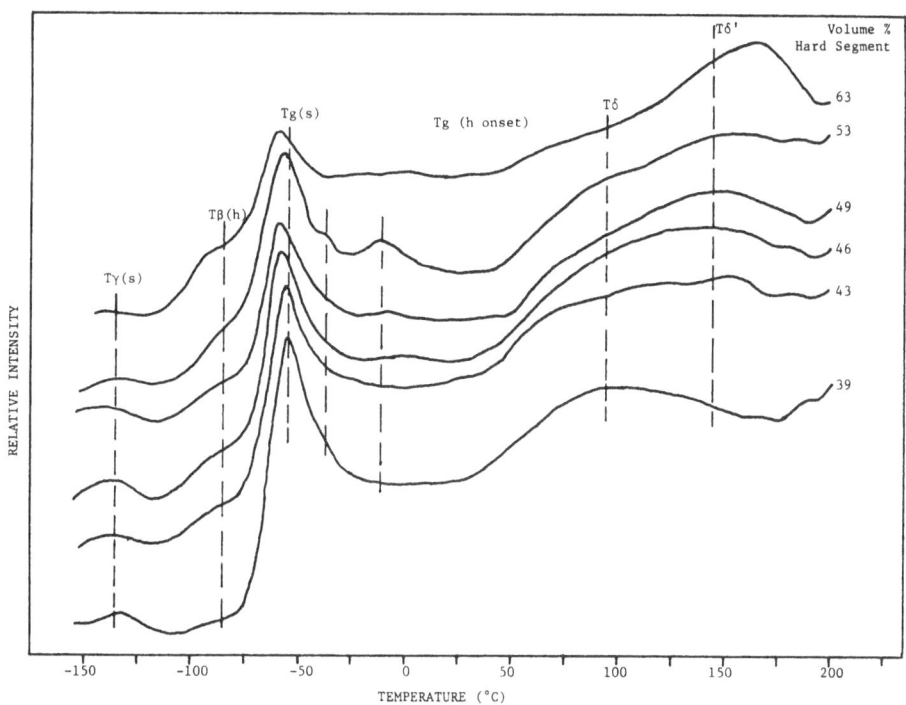

Figure 2. Plots of tangent delta versus temperature showing energy
loss peaks for the various segmented polyurethane formulations.

uous semi-crystalline or hard phase with some disruptions due to the
domains of the soft segment.

The major energy loss transition (Tg(s) which corresponds to a
glass transition of the soft phase occurs at a relatively low temperature at about -60°C. This finding is consistent with results
from other studies[1,2,3] which indicate that the Tg(s) loss peak
shifts to lower temperatures as the length of the soft segments is
increased. This lowering of the glass transition could indicate a
better ordered or larger domain developed by the lower molecular
weight polyethers. Thus a higher portion of the polyether soft segment will be well segregated from the aromatic hard segment urethane
linkages; as a consequence their motions will be less hindered. From
thermodynamic considerations, it also follows that phase segregation
is enhanced by increasing the molecular weight of a phase component.[5]
The glass transition of the soft segment actually decreases slightly
(-57.5°C to -60°C) with increasing hard segment concentration. This
result not only suggests that the hard segments are not penetrating
and mixing with the soft segments as the concentration ratios of the
phases are reversed, but that the soft segments are becoming slightly

better segregated from the hard segment phases.

Increased ordering of the phase domain structure of the hard segment domains in this series of samples appear to be supported by the
shifting and sharpening of the energy loss peaks assigned to mechanical motions of the hard segment as the MDI-EG concentrations are increased (see Table II). Other workers have reported glass transitions
of the hard segments ranging in temperature from 80°C to as high as
180°C.[6] In some instances, designation of these energy peaks as true
glass transitions of the hard segments have been justified on the
basis that dynamic mechanical spectra exhibit sharp, well-defined
second order transition peaks accompanied by rapid decreases in storage and loss moduli. In the present investigation, the location and
assignment of second order transitions to the hard segment microphase
is not straightforward.

The polymer containing the lowest hard segment content exhibits
a broad peak on the tangent delta curve defined by minima at 35°C and
about 160°C and a maximum at 95°C ($\delta$). As the hard segment is increased, a second broad peak appears to be growing with its maximum
at about 145°C ($\delta'$) while the first peak is diminishing. Indeed, the
area under this set of peaks on the tangent delta curve tends to increase with increasing concentration of hard segment chemical group-
ings (see Table III). Also, a concomitant decrease occurs in the
area under the glass transition peak of the soft segment.

Most workers agree that hydrogen bonding and chain packing are
the major factors that affect the structure stability of a hard segment domain. It seems reasonable that as the chemical reactants that
form hard segments are increased and microphase domain morphology
takes place in the polymer, there will be a certain distribution of
hard domains. At one extreme, only vaguely recognizable short-range

Table II. Energy Loss Transitions From DMS of Polyurethane Systems

| Sample Number | Volume, % Hard Segment | $T_G(s)$ | $T_\beta(H)$ | $T_\gamma(s)$ | $T_G(H)$ | $T_\delta$ |
|---|---|---|---|---|---|---|
| I | 39 | -57.5 | -85 | -132 | 30 | 95 |
| II | 43 | -58.0 | -82 | -135 | 34 | 125 |
| III | 46 | -59.0 | -85 | -137 | 39 | 135 |
| IV | 49 | -60.0 | -85 | -137 | 50 | 140 |
| V | 53 | -60.0 | -85 | -137 | 50 | 157 |
| VI | 63 | -62.0 | -90 | -138 | 50 | 165 |
| VII | 73 | -62.0 | -82 | -135 | 50 | 160 |

Table III.  Area Under Tg(s) Relaxations Versus Volume % Hard Segment

| SAMPLE NUMBER | VOLUME % HARD SEGMENT | *Tg(s) PEAK AREA |
|---|---|---|
| I | 39 | 0.110 |
| II | 43 | 0.076 |
| III | 46 | 0.061 |
| IV | 49 | 0.054 |
| V | 53 | 0.046 |
| VI | 63 | 0.025 |
| VII | 73 | |

*Tg(s) TAKEN FROM -25°C TO -75°C.

order will exist.  At the other extreme, an almost continuous hard phase characterized by long range order will prevail.  The degree and extent of order in domains are not easily measurable factors.  The tangent delta curve is further complicated by the fact that the Tg(h) transition, which triggers hard segment movements, is in turn swamped by the energy dissipated in the motion of these segments.

Even though we do not have examples of the extreme cases, the tangent delta curves (Figure 2) of the seven formulations tend to support wide variation in hard segment domain order.  The onset of the Tg(h) of the amorphous segments of the hard domain (Table II) occurs at about 30°C for the samples with the lowest concentration of hard segments and increases to about 50°C as the hard segment content is raised.  This change with temperature tends to support the argument that the low temperature peak (95°C) relaxation represents energy from motions of hard segments having domains of fairly limited short range order.  Gradual movement of the tangent delta maximum to higher temperatures along with increasing hard segment content suggests the presence of more and more domains containing long range order resulting in larger area under the relaxation.[7]

Two very low temperature energy loss transitions were recorded for this series of polymers.  Localized motion in polyether sequences has been reported to be responsible for the relaxation occurring at approximately -135°C.  McKnight and Kajayama have interpreted the peak to arise from the rotation of the methyl group in the propoxylated polyether chains.[8]

The hard segment should have an energy loss transition at a relatively low temperature analogous to those reported for polyesters

and polyamides.[9]  We tentatively assign the shoulder appearing on the
glass transition of the soft segment in the tangent delta curve to
the beta transition of the hard segment.  The assignment appears valid
since the peak heights increase with increasing hard segment concentration.

Differential Scanning Calorimetry Measurements

        Differential scanning calorimetry (DSC) studies on segmented
polyurethanes have been utilized in many studies.[1,3,10]  The major
conclusions from these workers were that the glass transition of the
soft segment, a thermal transition (Tg(h)) that occurs from approximately 30°C to 50°C and the melting point of the hard segment could
be determined by the use of DSC.  Figure 3 and Table V show the data
obtained from the samples using the DSC.  These results indicate that
as the volume percent of the hard segment increases, the change in
heat capacity at the glass transition decreases.  This is also evident
in the dynamic mechanical spectroscopy curves.  The second order transition that is observed between 30°C and 50°C can also be reflected in
the dynamic mechanical property data.  The melting endotherm of the
hard segment appears at approximately 220°C.  As the volume percent
hard segment increases, the heat of fusion for the melting of the hard
domain increases, but the melting temperature remains relatively constant.  This indicates that we are building a larger fraction of hard
domain in a lightly crosslinked polymer as more chain extender is
added.

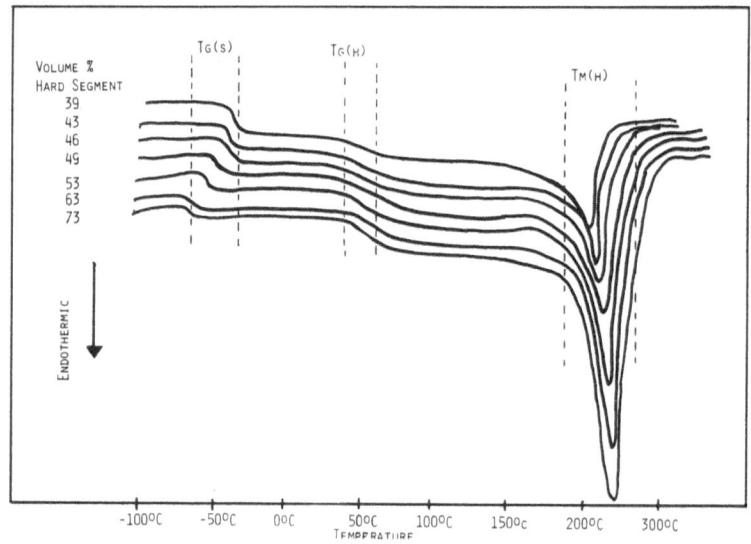

Figure 3.  DSC scans of polyurethane polymers at various hard
           segment levels.

## Physical Properties

Table IV shows the physical properties of the samples tested. As the volume percent hard segment increases, the flexural modulus, tensile and tear strength increase, and heat sag improves.  Also, as the volume percent hard segment increases, the elongation and Izod impact decrease.  This is consistent with the theory that as the

Table IV.  Physical Properties of Segmented Polyurethanes at Various Chain Extender Levels

| VOLUME % HARD SEGMENT | | 39 | 43 | 46 | 49 | 53 | 63 | 73 |
|---|---|---|---|---|---|---|---|---|
| FLEXURAL MODULUS | (PSI) | 27,000 | 43,000 | 67,000 | 93,000 | 110,000 | 160,000 | 227,000 |
| AT 23°C | (MPA) | 186 | 296 | 462 | 641 | 758 | 1102 | 1564 |
| TENSILE STRENGTH | (PSI) | 2800 | 3000 | 3400 | 3800 | 3300 | 4500 | 4600 |
| | (MPA) | 19 | 21 | 23 | 26 | 23 | 31 | 32 |
| ELONGATION | (%) | 235 | 210 | 190 | 125 | 75 | 63 | 53 |
| TEAR STRENGTH | (PLI) | 625 | 510 | 590 | 620 | 585 | 856 | 691 |
| | (KN/M) | 109 | 89 | 103 | 109 | 102 | 150 | 121 |
| IZOD IMPACT | (FT-LB/IN.) | N.A. | 8.7 | 9.4 | 6.0 | 4.1 | 3.5 | 2.4 |
| (NOTCHED) 23°C | (NM/M) | N.A. | 472 | 510 | 325 | 222 | 190 | 130 |
| HEAT SAG, 4" OVERHANG | | | | | | | | |
| AT 250°F/60 MIN.(IN.) | | 0.52 | 0.10 | 0.27 | 0.03 | 0.03 | 0.05 | 0.03 |
| (121°C) | (MM) | 13.20 | 2.54 | 6.86 | 0.76 | 0.76 | 1.27 | 0.76 |
| AT 325°F/30 MIN.(IN.) | | 3.20 | 3.28 | 1.05 | 0.53 | 0.44 | 0.50 | 0.45 |
| (162°C) | (MM) | 81.30 | 83.30 | 26.67 | 13.46 | 11.18 | 12.70 | 11.43 |
| SPECIFIC GRAVITY | (G/CC) | 0.985 | 0.950 | 1.00 | 1.03 | 0.962 | 1.09 | 0.943 |

Table V.  DSC Data From Various Chain Extender Levels in Polyurethane Polymers

| VOLUME % HARD SEGMENT | $T_G(S)$ °C | $T_G(H)$ °C | $T_M(H)$ °C | $\Delta H_F$ $T_M(H)$ JOULES/GRAM |
|---|---|---|---|---|
| 39 | -57.0 | 30 | 216 | 33.99 |
| 43 | -58.0 | 34 | 216.5 | 40.89 |
| 46 | -59.0 | 39 | 216.5 | 49.87 |
| 49 | -60.0 | 41 | 215.5 | 58.87 |
| 53 | -60.0 | 41 | 216.0 | 66.67 |
| 63 | -60.0 | 40 | 216.0 | 78.52 |
| 73 | -60.0 | 40 | 206.5 | 95.12 |

volume percent hard segment increases, so does the crystallinity in
the polymer. The change in density between the annealed and non-an-
nealed samples could be due to micropores being formed during annealing.

## Correlation of DMS and DSC Data

It is interesting to note that the DMS and the DSC can measure
the glass transition of the soft segment and possibly that of the
hard segment. The transition seen on the DMS at approximately 50°C
is also seen on the DSC. The DSC can measure the melting temperature
and the heat of fusion of the hard segment while the DMS reveals only
the onset of Tm.

## Effect of Hard Segment Volume

Figures 4, 5 and 6 show the flexural modulus, elongation and
Izod impacts of the samples vs. volume percent hard segment. It is
interesting to note that as the volume percent reaches approximately
50-55%, there is a leveling in the Izod impact curve and in the
elongation curve. The flexural modulus curve shows a non-linear increase. This indicates polymer morphological changes (possible phase
inversion) are occurring in the polymer at approximately 50-60 volume
percent. The DSC curve below 50 volume percent shows a sharp melting
endotherm for polymers that were not annealed before obtaining the
DSC spectra and the DSC spectra obtained on unannealed samples with

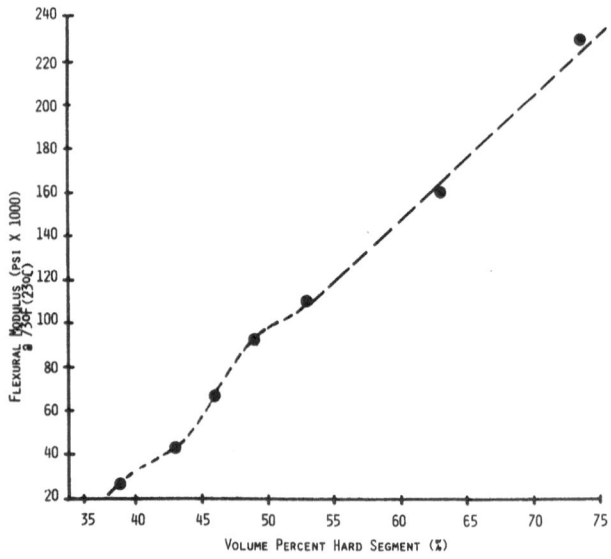

Figure 4.  Flexural modulus vs. volume percent hard segment.

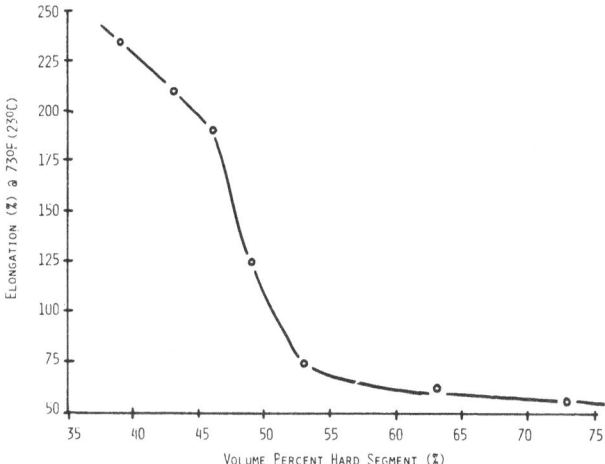

Figure 5.  Elongation vs. volume percent hard segment.

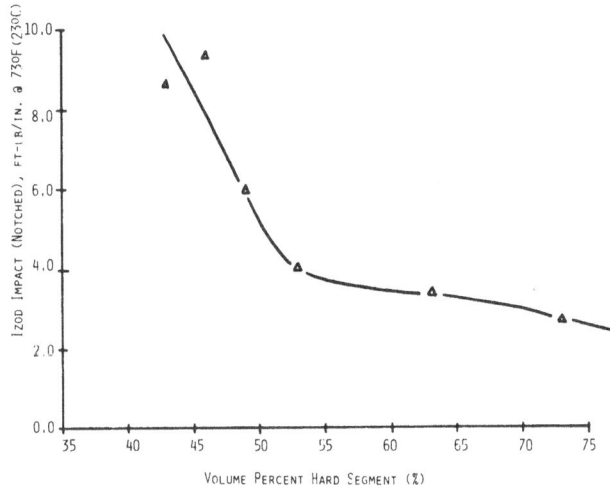

Figure 6.  Izod impact vs. volume hard segment.

volume percent hard segment above 50% show a broad multi-peak melting
point.  It is thought that this is due to microphase long range and
short range order in the polymer.  As can be seen from Figure 7, when
the polymer is annealed, the melting endotherm recorded by the DSC
sharpens until only one well defined peak appears.  This is consistent with the physical property data that shows that there is little
need for an annealing step below 50 volume percent at demold.  On the

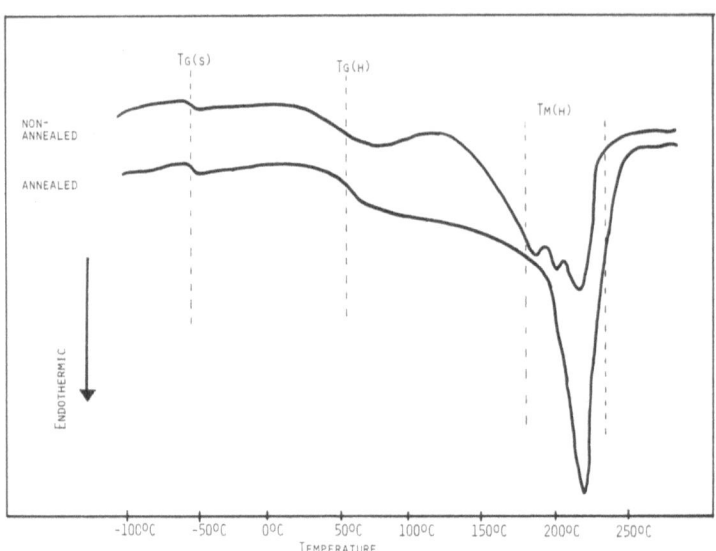

Figure 7.  DSC scans of the 63 volume percent hard segment polymer.

other hand, above 50 volume percent phase segregation of the polymer
is very poor and annealing is required to improve order in the hard
domain (see Table VI).  Since DMS spectra were run and the data obtained after the sample had stabilized approximately 3 minutes at
that temperature, we do not see this effect in the DMS data.  It is
well established that improved phase segregation improves physical
properties.[1,2,3,10]

The question of why good phase segregation is obtained below 50
volume percent and decreases rapidly above 50 volume percent hard segment in the polymer can now be answered.  Gellation and vitrification
are two microscopic phenomena which are encountered during the reaction which converts a liquid to a solid in the thermosetting process.[11]  Gellation is an example of a critical condition being reached
in a chemical reaction.  It is associated with a dramatic increase in
viscosity that occurs at a calculable degree of the reaction for the
particular reactive system.  On the molecular level, this corresponds
to the formation of crosslinked molecules of mathematically infinite
molecular weight.  Below the critical degree of reaction all of their
number and sizes increase with chemical conversion.  The mathematical
theory of gellation was formulated many years ago by Paul Flory.  Vitrification is the formation of a glassy solid.  In thermosets, this
usually follows gellation and then occurs as a consequence of the network becoming tighter through further chemical reaction (crosslinking).  A network structure will be a rubber or elastomer at a given
temperature if the segments between junction points of the network
are flexible.  If the segments are immobilized by gellation then they

Table VI.   Physical Properties Changes Due to Annealing
at Low and High Hard Segment Levels (160°C)

| | | 39 Volume % Hard Segment | | 73 Volume % Hard Segment | |
|---|---|---|---|---|---|
| | | Annealed | Non-Annealed | Annealed | Non-Annealed |
| Flexural | (psi) | 27,000 | 29,000 | 227,000 | 255,000 |
| Modulus | (MPa) | 186 | 199 | 1564 | 1757 |
| Tensile | (psi) | 2800 | 2820 | 4600 | 4300 |
| | (MPa) | 19 | 19 | 32 | 30 |
| Elongation | (%) | 235 | 210 | 53 | 13 |
| Tear Strength | (pli) | 625 | 587 | -- | -- |
| | (kN/m) | 109 | 103 | | |
| Izod Impact | (ft-lbs/in) | -- | -- | 2.40 | 1.10 |
| (Notched) | (Nm/m) | | | 130 | 60 |
| Heat Sag (250°F/60 min) 4" Overhang | | | | | |
| | (in) | 0.52 | 0.60 | 0.03 | 0.10 |
| (121°C) | (mm) | 13.20 | 15.2 | 0.76 | |
| Specific Gravity | | | | | |
| | (g/cc) | 0.985 | 1.00 | 0.949 | 0.952 |

are in a vitrified state.   Vitrification can prevent further reaction.
The overall transformation from liquid to gel to glass to rubber due
to chemical reaction is termed "cure".   It is conceivable that as the
volume percent hard segment increases, the polymer vitrifies before
gellation occurs thus immobilizing the softer segment.   As a result
phase segregation is decreased or interrupted entirely.   Conversely,
curing or annealing the samples above the glass transition of the hard
segment will allow the phase segregation of the soft and hard segment
to occur.   Annealing phenomenon increases the physical properties and
yields one melting endotherm on the DSC scan.   This insight into polymers above 50 volume percent hard phase reveals why the physical properties increase dramatically after annealing.   A correlation can be
made to polyethylene in which crystals are tied together by several
chains.   Such morphology is known to increase the physical properties.
So, below 50 volume percent hard segment gellation could occur before
vitrification and above 50 volume percent, vitrification could occur
before gellation is complete.


CONCLUSIONS

This study has shown that a hard segment domain can indeed be
obtained in a soft domain formed by a triol that lightly crosslinks
the polymer.   Good phase segregation is obtained below 50 volume percent of the hard segment; however, above 50 volume percent of hard
segment good phase segregation must be obtained through annealing of
the hard segments.   We have also seen indication that phase inversion
of the domains occurs at from 50-55% volume percent hard segment which

significantly affects the polymer properties.

## Acknowledgements

We wish to thank J. Butler for carrying out the DSC measurements, A. Pryor for assistance with the Rheometric Dynamic Mechanical measurements, and H. Helms for her work testing physical properties of the polymers. We want to thank L. C. Rubens for the helpful discussions.

## References

1.  R. J. Zdrahala, R. M. Gerkin, S. L. Hager and F. E. Critchfield, J. Appl. Polym. Sci., 24, 2041 (1979).
2.  R. Bonart, J. Macromol. Sci., B-2 (1), 115 (1968).
3.  S. L. Cooper and G. M. Estes, Multiphase Polymers, ACS, Ad. in Chem., 176 (1979).
4.  R. R. Aitken and G. M. F. Jeffs, Polymer, 18, 197 (1977).
5.  S. Krause, Macromolecules, 3, 84 (1970).
6.  J. W. C. Van Bogart, A. Liladnitkul and S. L. Cooper, Adv. in Chem., ACS, 176, 3 (1979).
7.  R. W. Seymour and S. L. Cooper, Macromolecules, 6, 48 (1973).
8.  T. Kajiyama and W. J. MacKnight, Macromolecules, 2, 254 (1969).
9.  F. Urabl, W. Sederal, J. M. Anderson and H. Hiltner, Polymer, 20, 51 (1979).
10. G. W. Miller and J. H. Saunders, ACS Polymer Preprints, 156 (1968).
11. J. K. Gillham, Polym. Eng. and Sci., 19, 10, 676 (1979).
12. J. L. Work, Macromolecules, 9, 759 (1976).
13. G. A. Senich and W. J. MacKnight, Adv. in Chem., 176, 97 (1979).
14. G. Kraus and K. W. Rollman, J. Polym. Sci., A-2 14, 1133 (1976).
15. J. Blackwell and K. W. Gardner, Polymer, 20, 13 (1979).
16. D. S. Huk and S. L. Cooper, Polym. Eng. and Sci., 11, 369 (1971).
17. D. J. Lyman, J. Heller and M. Barlow, Macromol. Chem., 84, 64 (1965).
18. L. L. Harrell, Macromolecules, 2, 6, 607 (1969).
19. J. L. Illinger and N. S. Schneider, Polym. Eng. and Sci., 12, 1, 25 (1972).
20. J. V. Dawking, Block Copolymers, 363 (1972).

# THE EFFECT OF ANNEALING ON THE THERMAL

# PROPERTIES OF RIM URETHANE ELASTOMERS

Richard J. G. Dominguez

Texaco Chemical Co.
P. O. Box 15730
Austin, TX 78761

## INTRODUCTION

Reaction Injection Molded (RIM) polyurethane elastomers are receiving a good deal of attention from the automobile industry. In
recent papers by Liedtke[1] at General Motors Manufacturing and Development, Mikulec[2] at Ford PDAO and Lloyd[3] at Texaco, it has become apparent that RIM polyurethane elastomers are a leading candidate for
making automobile exterior body panels. One very important performance property in this end use is thermal dimensional stability.
This is so for two reasons: first, paint bake cycles tend to involve
relatively high temperatures, around 280-350°F and second, in use
temperatures are often rather high for long periods of time. A part
made from a RIM polyurethane elastomer should be able to withstand
these conditions without an unacceptable degree of distortion.

It is well known that the properties of thermoplastic polyurethane (TPU) elastomers are dependent on thermal history. Workers
such as Cooper[4], Jacques[5] and Wilkes[6] have demonstrated that different annealing conditions can change the thermal profile of TPU elastomers as seen by differential scanning calorimetry (DSC). In fact,
DSC has emerged as one of the most important methods for studying
the thermal properties of plastics materials[7].

In order to understand the origin of the unique properties of
TPU elastomers, let us briefly comment on the nature of these materials. Polyurethane elastomers are block copolymers, specifically
$(AB)_n$ block copolymers.[8] Three basic ingredients make up a polyurethane elastomer: (a) an isocyanate, (b) a chain extender and
(c) a macroglycol. The isocyanate is usually an aromatic diisocyanate, the chain extender is usually a low molecular weight diol such

as ethylene glycol or 1,4-butanediol and the macroglycol is usually
a polyether or polyester hydroxyl terminated resin.  The isocyanate
and chain extender react to form the hard segment and the preformed
macroglycol inserts itself into the hard segment as it polymerizes,
providing the soft segment.  The hard segment provides stiffness and
high temperature performance properties while the soft segment provides toughness to the material.  As is the case for most polymer
pairs, polyurethane block copolymers tend to phase separate.  The
morphology of the resulting material is best described as discernible
regions primarily composed of hard segments in juxtaposition to discernible regions primarily composed of soft segments.

In a previous paper[9] the effect of annealing on the properties
of several thermoset RIM polyurethane elastomers has been discussed.
It is clear from this work that thermal dimensional stability is
dramatically improved when these materials are annealed at temperatures of 300°F or greater.  It was proposed that the observed improvement in this important property results from (a) hard segment ordering, i.e., crystallization and (b) morphological changes, i.e., phase
separation or changes in phase structure.

In this paper we will attempt to provide evidence for these
mechanisms in thermoset polyurethanes and to describe them in greater
detail.  We have relied on DSC, thermal mechanical analysis (TMA),
wide angle X-ray diffraction (WAXD) and small angle X-ray scattering
(SAXS) to supply the evidence for the suggested mechanisms.

## Materials

The following four basic materials were studied.  These materials have room temperature flexural modulus values of about 140,000
psi.  In all cases a high molecular weight, high primary hydroxyl
polyether macroglycol is employed for the polyol.

Formula A:  Polyol/ethylene glycol/liquid MDI

Formula B:  Polyol/ethylene glycol/MDI-enriched
            polymeric isocyanate

Formula C:  Polyol/ethylene glycol/polymeric isocyanate

Formula D:  Polyol/1,4-butanediol/liquid MDI

In addition to the above formulations, a fifth material of
similar composition of Formula A but with a room temperature flexural
modulus of about 90,000 psi was also studied.  In this paper it is
called Formula E.

## Sample Preparation

The results presented in this paper were obtained from plaques
molded in a Cincinnati-Milacron LRM-II RIM machine.  Fifteen minutes
after release from the mold, the plaques were annealed in forced
draft ovens for thirty minutes.  Appropriate test samples were then
cut from the plaques and allowed to equilibrate at standard ASTM conditions for at least one week before testing.


## DISCUSSION

## High Temperature Performance; The Heat Sag Test

One method of rating the relative thermal dimensional stability
of RIM polyurethane elastomers is the heat sag test.  In this test,
one end of a sample of given dimensions is clamped onto a fixture as
shown in Fig. 1.  A specified length of the sample extends unsupported from the fixture in the horizontal plane.  The vertical distance from the unsupported end of the sample to the base of the test
fixture is measured before and after subjecting the sample to a specific temperature for a given time.  Measurements on a one-inch wide
one-eighth inch thick sample having a 6-inch unsupported overhang and
heated at 325°F for 30 minutes give consistent and discriminating
results.

In Fig. 2, the 325°F, 6-inch heat sags for Formulas A, B, C and
D are given as a function of annealing temperature.  The heat sag
drops dramatically as the annealing temperature increases to 300°F.
Thereafter, the heat sag remains relatively constant.  The remainder
of this paper will be devoted to understanding the nature of this
improvement in thermal dimensional stability.

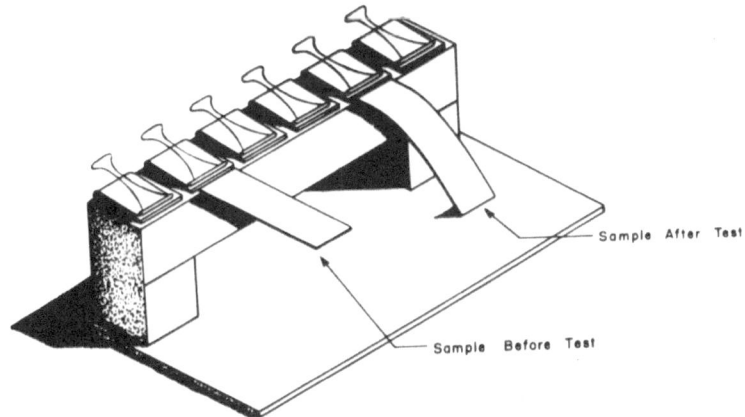

Figure 1.  The heat sag test.

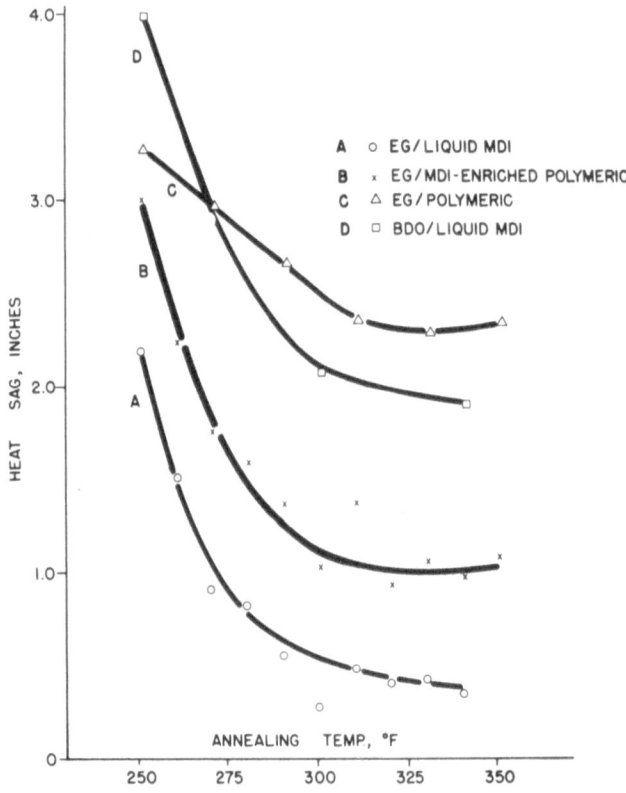

Figure 2.   The heat sag (325°F, 30 min., 6" overhang) of the Formula A, B, C and D elastomers as a function of annealing
tempeature.

## DSC Measurements

We have studied the thermal response of these materials quite
extensively by the DSC technique.  In Fig. 3 the DSC profile of the
Formula A elastomer is given as a function of different annealing
conditions.  This material contains the ethylene glycol/liquid MDI
hard segment.  Unannealed, the material possesses quite a bit of thermal activity between 45 to 145°C.  When annealed at 250°F or 340°F,
the activity virtually disappears.  In order to decide what causes
this activity, we investigated the DSC of the ethylene glycol/liquid
MDI hard segment by itself.  This was accomplished by synthesizing
the polymer in THF and annealing part of this product at 340°F.  The
DSC data for the hard segment only system is given in Fig. 4.  We
see that in the unannealed state, similar thermal activity between
45 and 140°C is present as is the case in the analogous elastomer
(Fig. 3A).  On annealing at 340°F these features virtually disappear.
The synthesis of this hard segment in THF reproduces rather well an

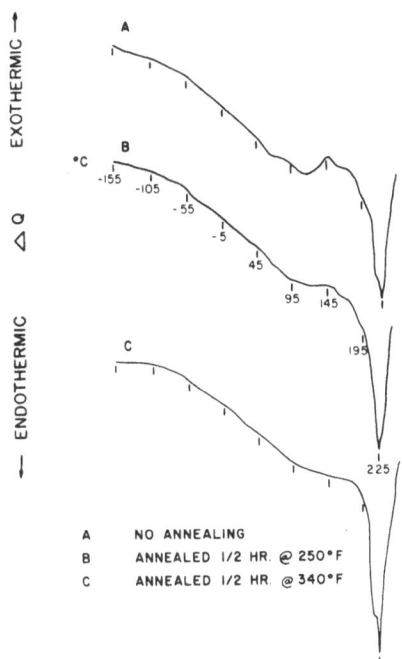

Figure 3.   The DSC of the Formula A elastomer as a function of
            annealing temperature.

important feature of the RIM process.  The monomers polymerize to a
certain point and suddenly the polymer precipitates from solution.
The suddenness of this precipitation is analogous to what occurs when
a RIM polyurethane elastomer forms, namely, a rapid polymerization
followed by a sudden solidification from liquid to melt to solid elastomer.  This is in sharp contrast to how thermoplastic polyurethanes
are synthesized.  The relatively long time factor present in the synthesis of TPU elastomers allows the material to order and develop a
relatively thermodynamically stable configuration.  Thermoset polyurethane elastomers made by the RIM process have a much greater probability to be frozen in a relatively metastable condition.  The DSC
data in Fig. 4 lead one to believe that the thermal activity observed
in the unannealed RIM elastomer (Fig. 3A) is related only to the hard
segment.

At this point it is appropriate that we turn our attention to the
question of whether or not significant chemical reactions occur during
annealing.  Clearly, the DSC data discussed to this point could be
accounted for by proposing that it results from further chemical reaction.  In Fig. 5, the Fourier Transform infrared (FTIR) of the ethylene glycol/liquid MDI hard segment is given for the region 1900 to
2500 cm$^{-1}$.  Free isocyanate absorbs at about 2100 cm$^{-1}$ in the IR.

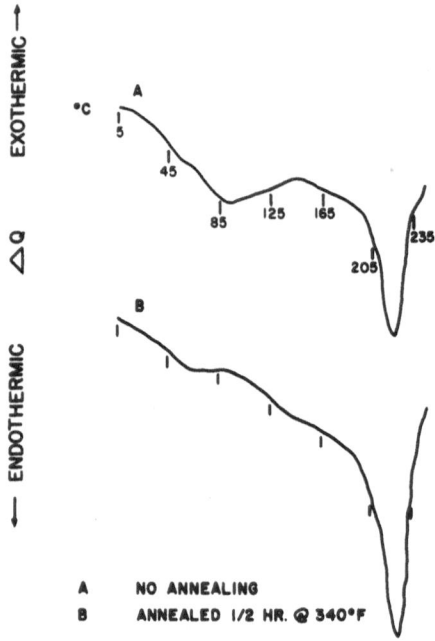

Figure 4.    The DSC of the ethylene glycol/liquid MDI hard segment
as a function of annealing temperature.

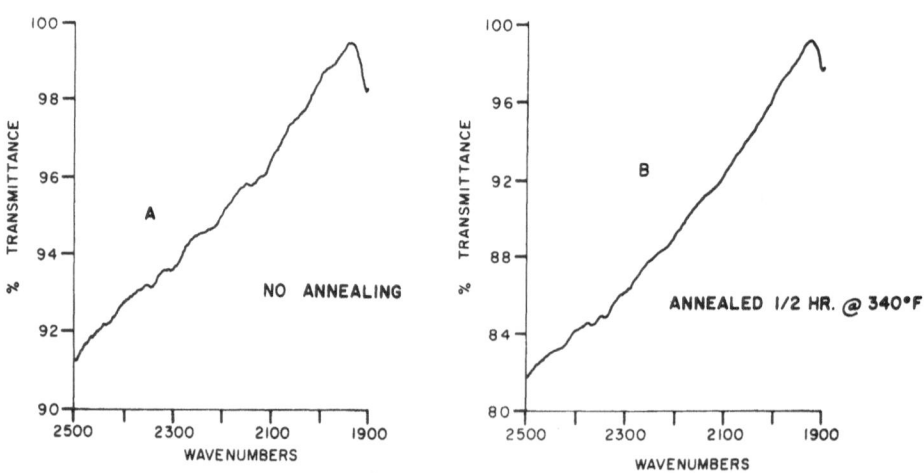

Figure 5.    The FTIR of the ethylene glycol/liquid MDI hard segment
as a function of annealing temperature.

This band has a relatively high molar extinction coefficient. It is evident from Fig. 5 that there is no appreciable difference in the free isocyanate content of the material before and after annealing. In fact, even in the unannealed hard segment, the presence of only a trace level of free isocyanate is suggested by the spectra. It, therefore, seems reasonable to account for the changes on annealing observed in the DSC of both the elastomer and the hard segment only systems (Figs. 3 and 4) on the basis of physical processes such as hard segment ordering. We will present WAXD data later that tend to confirm this hypothesis.

Two other features of the DSC profiles in Fig. 3 are worth noting. At the low temperature end there is a very weak transition around -50°C. This is the glass transition (Tg) of the polyether soft segment. At the high temperature end there is a rather pronounced transition around 210°C. This is the melt endotherm (Tm) of the hard segment. This two-phase feature occurs in the DSC of all the elastomers described in this paper.

In Fig. 6, the DSC profile of the Formula B elastomer is given as a function of different annealing conditions. This material contains the ethylene glycol/MDI-enriched polymeric isocyanate hard segment. Note that unannealed, the poor definition of the hard segment melt endotherm indicates that it does not possess much crystallinity

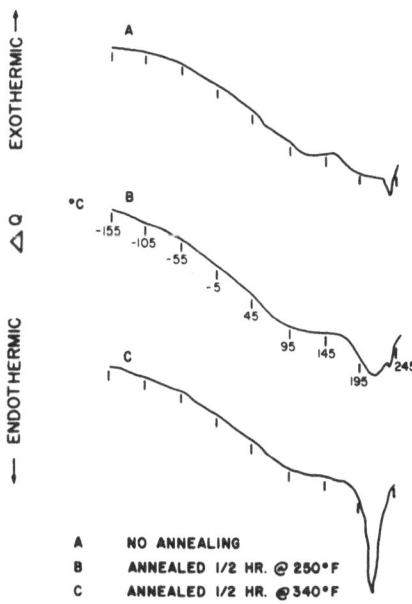

A    NO ANNEALING
B    ANNEALED 1/2 HR. @ 250°F
C    ANNEALED 1/2 HR. @ 340°F

Figure 6.   The DSC of the Formula B elastomer as a function of annealing temperature.

(Fig. 6A) in contrast to the significant crystallinity observed for
the unannealed ethylene glycol/liquid MDI material (Fig. 3A).  This
probably occurs because the MDI-enriched polymeric isocyanate intro-
duces a significant amount of trifunctional and higher functionality
isocyanate into the hard segment.  This will result in crosslinking
with the natural consequence of less ordered hard segments leading
to lower crystallinity.  However, when the material is annealed at
250°F, some crystallinity develops (Fig. 6B).  When annealed at 340°F,
the DSC exhibits a rather well-defined crystalline melt endotherm
(Fig. 6C).  Apparently, annealing softens amorphous hard segment regions and encourages the formation of crystallites which we then see
melting in the DSC of the annealed material.

In Fig. 7, the DSC profile of the Formula C elastomer is given
as a function of different annealing conditions.  This material contains the ethylene glycol/polymeric isocyanate hard segment.  The
polymeric isocyanate is even higher in functionality than the MDI-
enriched polymeric isocyanate, so crosslinking within the hard segment is relatively high in this system.  Consequently, not only the
unannealed material (Fig. 7A) but even when the material is annealed
at 340°F (Fig. 7C) a well-defined melt endotherm does not develop.
True, the hard segment melt endotherm develops better definition as

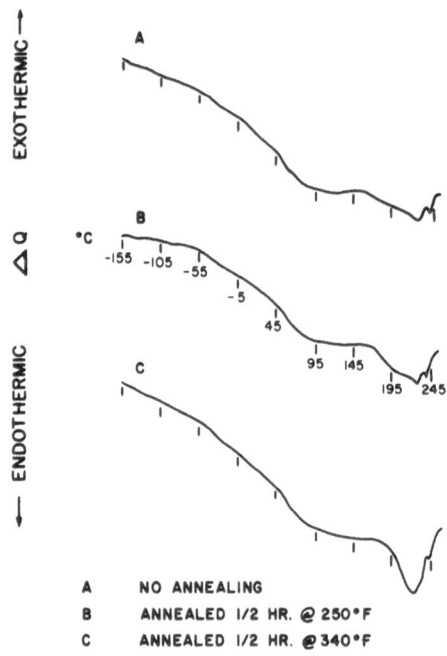

A       NO ANNEALING
B       ANNEALED 1/2 HR. @ 250°F
C       ANNEALED 1/2 HR. @ 340°F

Figure 7.   The DSC of the Formula C elastomer as a function of
            annealing temperature.

the material is annealed at higher temperature; however, it never achieves the same definition as the hard segments of Formulas A & B (Figs. 3 and 6).

In Fig. 8, the DSC profile of the Formula D elastomer is given as a function of different annealing conditions. This material contains the 1,4-butanediol/liquid MDI hard segment. The DSC data show dramatic changes in the hard segment melting region. The results strongly suggest that annealing at higher temperatures causes better developed order within the hard segment.

These results correlate rather well with the heat sag data for these materials shown in Fig. 2. As one anneals material made from Formulas B, C & D at higher temperatures, the heat sag of the materials is dramatically reduced. In the previous discussion, we have shown that the DSC profiles of these three materials show that as one anneals them at higher temperatures, the hard segment tends towards a greater degree of order. However, it should be pointed out that if the materials are annealed at temperatures within the hard segment melt endotherm, a total collapse of hard segment order re-

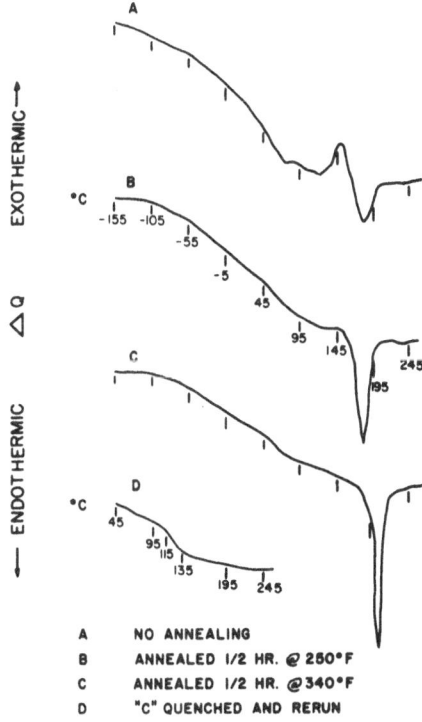

A    NO ANNEALING
B    ANNEALED 1/2 HR. @ 250°F
C    ANNEALED 1/2 HR. @340°F
D    "C" QUENCHED AND RERUN

Figure 8.  The DSC of the Formula D elastomer as a function of annealing temperature.

sults. This is illustrated in Fig. 8D. Here, the elastomer of Fig.
8C was heated in the DSC instrument to a temperature just beyond the
melt endotherm peak and then quenched. When the DSC trace is rerun,
we observe a well-defined hard segment Tg with no evidence of the
hard segment melt endotherm observed in Fig. 8C. The hard segment
becomes completely disordered into an amorphous glass if the annealing temperature is within the hard segment melt endotherm. Similar
results are observed in the other RIM elastomers studied.

In the ethylene glycol/liquid MDI system (Formula A) we observed
changes in the DSC profiles of unannealed versus annealed material
in the region 45° to 145°C. However, we did not observe any significant changes in the hard segment melt endotherm. We will devote a
good deal of attention to this question later on in this paper.

## Wide Angle X-ray Diffraction Studies

In Table I, the WAXD results of the ethylene glycol/liquid MDI
hard segment and three elastomers (Formulas A, B and D) are given as
a function of annealing temperature. In all cases, the diffraction
peak half width decreases with a consequent increase in average crystallite thickness. These data indicate that in all these cases, the
material annealed at 340°F is more ordered than the unannealed material. The data are further evidence that hard segment ordering is a
reasonable explanation for some of the changes observed in the DSC
profiles of these materials as a function of annealing temperature.

Table I.  WAXD of the Ethylene Glycol/Liquid MDI Hard Segment and
          the Formula A, B and D Elastomers as a Function of
          Annealing Temperature.

| | D-SPACING (Å) | HALF-WIDTH (degrees) | AVERAGE CRYSTALLITE THICKNESS (Å) |
|---|---|---|---|
| **EG/143L HARD SEGMENT** | | | |
| No Annealing | 4.35 | 10.4 | 1.35 |
| Annealed 1/2 Hr. @ 340°F | 4.40 | 8.0 | 1.68 |
| **EG/143L/POLYOL** | | | |
| No Annealing | 4.48 | 9.0 | 1.56 |
| Annealed 1/2 Hr. @ 340°F | 4.48 | 8.5 | 1.66 |
| **EG/191/POLYOL** | | | |
| No Annealing | 4.48 | 8.8 | 1.60 |
| Annealed 1/2 Hr. @ 340°F | 4.39 | 8.0 | 1.76 |
| **1,4 BDO/143L/POLYOL** | | | |
| No Annealing | 4.53 | 9.0 | 1.56 |
| Annealed 1/2 Hr. @ 340°F | 4.62 | 8.5 | 1.65 |

## The Formula E System

In Fig. 9, the DSC profile of the Formula E system is given as a function of different annealing conditions. This elastomer is similar to the elastomer of Formula A except that it has less hard segment and consequently more soft segment. The same features are observed here as were observed in Fig. 3. Note that due to the increased soft segment concentration of this system, we can see the Tg of the soft segment more clearly (about -50°C). The soft segment Tg gets more pronounced and occurs at lower temperature as the annealing temperature increases. In fact, the Tg decreases from about -47°C unannealed to about -54°C annealed at 325°F. This is seen more clearly in Fig. 10, which is a TMA study of the Formula E system as a function of annealing temperature. Note the pronounced penetration (Fig. 10B and $B_1$) observed in the annealed material near the soft segment Tg. In the unannealed material (Fig. 10A and $A_1$) the penetration is hardly noticeable. The above discussion suggests that the material becomes more phase separated as the annealing temperature is increased.

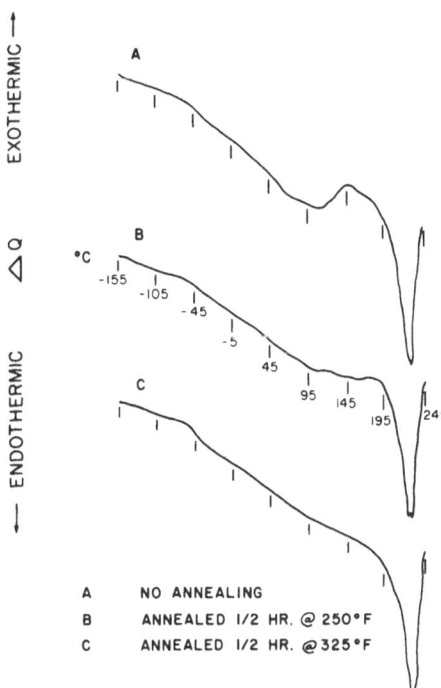

Figure 9.   The DSC of the Formula E elastomer as a function of annealing temperature.

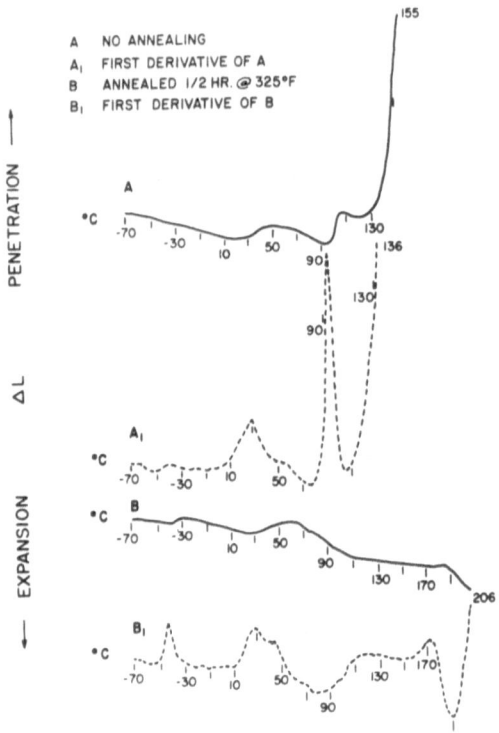

Figure 10.    The TMA of the Formula E elastomer as a function of
              annealing temperature.

We have also studied the wide angle X-ray diffraction of the
Formula E system as a function of annealing temperature.   These data
are reported in Fig. 11.   Clearly, the unannealed material (Fig. 11A)
is less crystalline than the annealed materials (Fig. 11B and C) since
the diffraction rings are more clearly pronounced in the annealed materials than in the unannealed material.   Comparing the material annealed at 250°F (Fig. 11B) with the material annealed at 325°F (Fig.
11C) it is seen that the former material is more crystalline than the
latter.   When we commented on the DSC of the Formula A system (Fig. 3)
we noted the differences between unannealed and annealed material near
the hard segment melt endotherm region (45 to 145°C) (Fig. 3A versus
Fig. 3B and C).   We observed that there were no significant differences in the hard segment melt endotherm (around 210°C) of the material  annealed at 250°F (Fig. 3B) or annealed at 340°F (Fig. 3C).
The same qualitative observations hold for the DSC profiles of the
Formula E system (Fig. 9).   We have made quantitative measurements
on the enthalpy of hard segment melting (Fig. 12) for the Formula E
system and find that the highest value is obtained for the material
annealed at 250°F, next highest for the material annealed at 325°F
and least for the unannealed material.   These data correlate rather

A                          B                          C

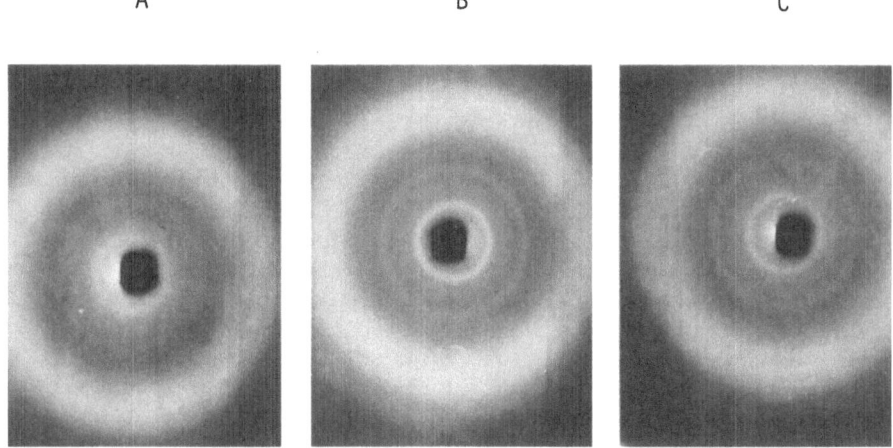

A     NO ANNEALING
B     ANNEALED @ 250°F FOR 1/2 HR.
C     ANNEALED @ 325°F FOR 1/2 HR.

Figure 11.   The WAXD of the Formula E elastomer as a function of
             annealing temperature.  (A) No annealing; (B) annealed
             at 250°F for 1/2 hr.; (C) annealed at 325°F for 1/2 hr.

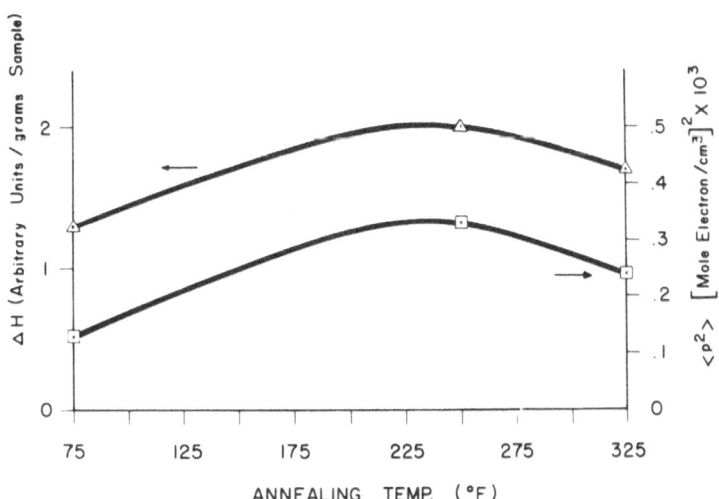

Figure 12.   $\Delta H$ from the DSC and $<\rho^2>$ from the SAXS of the Formula
             E elastomer as a function of annealing temperature.

well with the wide angle X-ray diffraction data in Fig. 11.

 We have also studied the SAXS of the Formula E system as a function of annealing temperature (Fig. 13). Note that as the annealing temperature is increased a better defined shoulder at small angles tends to emerge. The emergence of this shoulder can occur as a result of various mechanisms: (a) increased hard segment crystallinity, (b) greater phase separation or (c) changes in phase continuity possibly as a result of the hard segment phase becoming more continuous and the soft segment phase becoming more discontinuous. Mean square electron density fluctuation $\langle \rho^2 \rangle$ data can be obtained from the SAXS data (Fig. 12). We have calculated this parameter as a function of annealing temperature and find that it follows the same trend as the enthalpy measurements. Thus, the highest value of $\langle \rho^2 \rangle$ is obtained for the material annealed at 250°F, next highest when annealed at 325°F and least for unannealed material. This parameter usually indicates the degree of phase separation or crystallinity of a material. From the preceding discussion of WAXD and enthalpy data, it seems reasonable to assign the behavior of $\langle \rho^2 \rangle$ to changes in crystallinity. The WAXD data, enthalpy measurements and $\langle \rho^2 \rangle$ parameter as a function of annealing temperature tend to discredit interpreting the low angle shoulder observed in the SAXS (Fig. 13) on the basis of increased crystallinity since the former three methods indicate less crystallinity in the material annealed at 325°F vs. the material annealed at 250°F. In contrast, the low angle SAXS shoulder is more pronounced in the material annealed at 325°F vs. the material annealed at 250°F. A more reasonable explanation of Fig. 13 is to attribute the emergence of the shoulder at small angles to increased phase separation and/or changes in phase continuity.

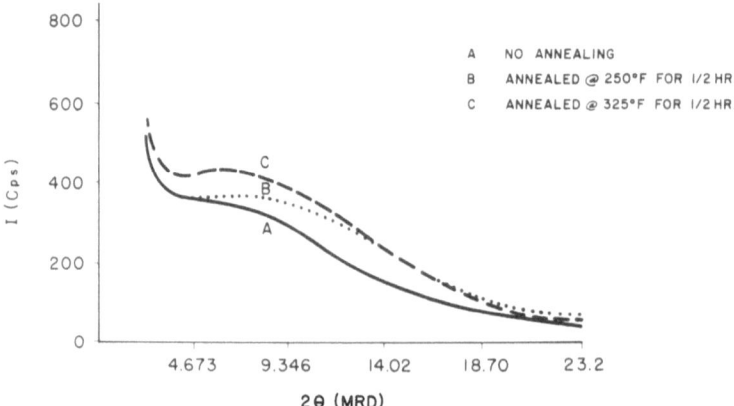

Figure 13.   The SAXS of the Formula E elastomer as a function of
             annealing temperature.

The ethylene glycol/liquid MDI hard segment elastomer systems
we have investigated (Formulas A and E) have given us a unique opportunity. The thermal dimensional stability (heat sag) of both elastomers dramatically improves as a function of increasing annealing
temperatures. However, the extent to which hard segment crystallinity
is involved in this property is slight in comparison to the other systems discussed in this paper. In fact, hard segment crystallinity is
present to a lesser degree when the Formula E system is annealed at
elevated temperatures (325°F) than when it is annealed at lower temperatures (250°F). The SAXS work on the Formula E system indicate
that there are significant morphological changes which occur within
this material as a function of annealing temperature. The ethylene
glycol/liquid MDI hard segment elastomers have thus provided us with
a system where the relation between these morphological changes to
heat sag is not masked by dramatic crystallinity changes. In Table
II, temperature and strength properties for the Formula E system are
given as a function of different annealing conditions. As in all the
other RIM polyurethane elastomers studied, thermal dimensional stability (heat sag) dramatically improves with increasing annealing temperature. However, when comparing material annealed at 250°F or
325°F, little change in strength properties is observed. When the
unannealed material is compared to the annealed materials, strength
properties do significantly improve. This is possibly due to some
macroglycol-isocyanate chemical reaction occurring during the annealing step.

Table II.  Properties of the Formula E Elastomer as a Function
           of Annealing Temperature

| | NO ANNEALING | ANNEALED 1/2 HR. @ 250°F | ANNEALED 1/2 HR. @ 325°F |
|---|---|---|---|
| Ultimate Tensile Strength; psi | 3800 | 4400 | 4400 |
| Ultimate Elongation; % | 110 | 145 | 145 |
| Flexural Modulus ; psi Measured At -20°F | 204,000 | 196,000 | 173,000 |
| 77°F | 97,600 | 96,000 | 97,700 |
| 158°F | 47,800 | 54,400 | 60,600 |
| 325°F | 21,100 | 21,900 | 23,000 |
| Heat Sag; inches 1/2 Hr. @ 325°F, 6" Overhang | >4 | 3 | 0.4 |
| IZOD Impact, Notched; ft.lbs./in. | 11 | 13 | 11 |

CONCLUSIONS

This stuay nas snown that thermoset RIM polyurethane elastomers
dramatically improve in thermal dimensional stability (heat sag) when
annealed at higher temperatures (300 to 350°F). Data have been presented which supply evidence for the occurrence of three different
structural changes when RIM polyurethane elastomers are annealed at
increasing temperatures, all of which should result in better thermal
dimensional stability. First, in three of the elastomers hard segment
ordering (crystallinity) increases as a function of increased annealing temperature; second, phase continuity changes, i.e., a more continuous hard segment phase and a less continuous soft segment phase
possibly occur with increased annealing temperature; and third, more
complete phase separation appears to occur with increased annealing
temperature.

## Acknowledgement

The author wishes to express his appreciation to G. L. Wilkes
and S. Abouzahr for conducting the SAXS and some of the WAXD and ΔH
measurements and to G. P. Speranza, R. F. Lloyd and D. M. Rice for
their assistance and cooperation in the preparation and review of
this manuscript. Also to Texaco Chemical Company for their support
of this project.

## Nomenclature

    Liquid MDI = Isonate 143L*
    MDI-enriched polymeric isocyanate = Isonate 191*
    Polymeric isocyanate = PAPI 901*


## References

1.  M. W. Liedtke, J. Cell. Plast., March/April, 102 (1978).
2.  M. J. Mikulec, 34th Annual Technical Conference, Reinforced
       Plastics/Composites Institute, SPI, January/February, Section
       11B, (1979).
3.  R. F. Lloyd, SAE Congress and Exposition, Paper No. 800514,
       Society of Automotive Engineers, (Feb. 29, 1980).
4.  R. W. Seymour and S. L. Cooper, Macromolecules, 6, 48 (1973).
5.  C. H. M. Jacques, "Effects of Annealing on the Morphology and
       Properties of Thermoplastic Polyurethanes," in "Polymer
       Alloys: Blends, Blocks, Grafts and Interpenetrating Net-
       works," p. 287, D. Klempner and K. C. Frisch, Eds., Plenum
       Press, New York (1977).

---

*Products of the Upjohn Chemical Company
  E10

6.  Y.-J. P. Chang and G. L. Wilkes, J. Polym. Sci., Polymer
       Physics Ed., 13, 455 (1975).
7.  M. J. Richardson and N. G. Savill, Brit. Polym. J., 11, 123
       (1979).
8.  S. L. Cooper, J. C. West and R. W. Seymour, "Polyurethane Block
       Polymers," in :Encyclopedia of Polymer Science and Technology,"
       Volume 1 Supplement, p. 521, H. F. Mark and N. M. Bikales,
       Eds., Interscience Publishers, Div. of John Wiley and Sons,
       Inc., New York (1976).
9.  D. M. Rice and R. J. G. Dominguez, Polym. Eng. Sci., 20, 1192
       (1980).

# EXPERIMENTAL STUDIES OF PHASE SEPARATION IN REACTION

# INJECTION MOLDED (RIM) POLYURETHANES

R. E. Camargo, C. W. Macosko, M. V. Tirrell
and S. T. Wellinghoff

Department of Chemical Engineering
and Materials Science
University of Minnesota
Minneapolis, MN 55455

## INTRODUCTION

Reaction injection molding or RIM is the high speed production
of polymer parts directly from low viscosity reactants injected into
a mold.  In contrast to thermoplastic injection molding, where the
part solidified upon cooling, in RIM the shape is set by fast (in-
situ) polymerization.  The reaction leading to the formation of
the polymer can be initiated by mixing or by heat transfer.[1]  For
the mixing activated process two highly reactive monomers or prepolymers, in stoichiometric ratios, are brought into intimate contact by impingement mixing.  The thermally activated RIM materials,
on the other hand, have to be initiated by an increase in monomer
temperature.[1]

Necessary to the RIM process, thus, are high reactivity and absence (or ease of removal) of by-products during the polymerization.
To date, the major commercial RIM processed materials are polyurethanes.  The reasons for this are that polyurethane chemistry has
been able to provide in addition to the requirements already mentioned, a desirable variety of mechanical properties through the
wide range of intermediates, polyols, isocyanates and extenders which
can be employed.[2-4]

### Measurement of Phase Separation in Segmented Polyurethanes

The goal of this section is not to present a detailed review
of the extensive literature in the area, but to provide the reader
with the major advantages of the techniques used in the present work.

Excellent reviews are available from the literature.[4,5,11]

Polyurethane elastomers are multiblock copolymers of the type
$\{\text{AB}\}_n$.[5] The hard segments are immobile regions made from a step
growth polymerization of diisocyanates and low molecular weight diamines or diols. These regions are interconnected by polyether or
polyester based soft segments which provide the flexible character
to the polymers.

In commercial RIM systems, these hard and soft segments have
a slight polyfunctionality, yielding a low degree of crosslinking.
However, the physical crosslinking of the polyether chains provided
by the associated hard segments predominantly controls the mechanical properties. Thus, the relationship between processing conditions and structure/properties can be better studied using linear
systems which have greater characterization possibilities because
of their solubility.[2,6]

The polyurethane reaction is highly exothermic and polyurethanes formed by RIM can develop spatial inhomogeneities due to
temperature gradients across the mold which may be as high as
100°C.[2,6,7] These differing thermal histories yield differences in
molecular weight[2,6] and more complex changes in morphology and crystal structure as one transverses the mold from surface to center.[6,7]

Previous studies in our group and others[2,8-10,41,42] have been
directed to follow the development of heterogeneous structures during the polymerization. Monitoring of the reaction in slow systems
by light transmission,[8] differential scanning calorimetry[10] and infrared spectroscopy[9] has shown that the onset of phase separation
takes place at a fairly low number average sequence length of hard
segments. Mixing studies by Kolodziej et al[41-42] have demonstrated
that the degree of phase separation is related to the quality of
mixing. Low degree of mixing favors the formation of unattached
hard segments probably due to a diffusion limited polymerization.
Thomas and co-workers[7] have also presented evidence to suggest that
some heterogeneities in RIM processed urethanes may appear during
the mixing step.

In any case it is evident that the final properties and structure in these systems are largely determined by the polymerization
conditions. Differences in thermal history, reaction rate or mixing may give rise to quite different morphologies and sample properties.

Dynamic mechanical property studies have been successful in
unraveling the phase behavior of polyurethanes.[12-19] In 1966 Cooper
and Tobolsky[12] proposed that the particular viscoelastic behavior
of some polyurethane systems was caused by the association of dif-

ferent blocks in separate domains. In general it has been recognized that the extent of mixing of hard and soft segments may be estimated from changes in the glass transition temperature of the respective blocks.

Two separate Tg's are observed for a solid completely phase separated into hard and soft regions while increased phase inter-mixing results in a gradual merging of the Tg's at some intermediate temperature. These interactions can be also evaluated from broadening in the relaxation peaks and a less defined rubbery plateau of the elastic modulus.

Wide angle X-ray scattering has been used by several investigators to determine the structure of crystalline hard segments.[6,20-25] Several models have been proposed for the hard segment packing in MDI based systems.[22,24] The major hard segment crystalline reflections are well established[23] and the appearance of these in a given polyurethane sample can be interpreted in terms of hard segment purity and organization, in particular when no soft segment crystallinity is possible. It is for this reason that we have chosen WAXS for our particular study.

Hydrogen bonding is an essential part in the stabilization of hard segment structure, and has been taken into account in different hard segment models.[22,24] The extent of interurethane hydrogen bonding can be estimated by infrared spectroscopy.[9,11,26-34] In urethanes, a large fraction (ca. 80%) of the NH groups of the urethane linkage (- N - C - O -) is hydrogen bonded. The degree of mixing
$\qquad$ H $\quad$ O
between the phases is measured by considering the various groups capable of associating with the NH groups. In polyether systems, the C = O of the urethane and ether oxygens are major hydrogen bonding acceptors.[11,26,28,30,31] Since these two groups are preferentially located in different phases (assuming that little or no association takes place between urethane linkages in the soft phase), comparison of the fraction of hydrogen bonded C = O groups to that of the NH groups which is always higher can give an idea of the extent of urethane-soft segment association.

Cooper and co-workers[11,28,33] have applied this analysis to several systems based on MDI/BDO hard segments and polyether soft segments and have found that the difference between the fraction of NH and C = O hydrogen bonding correlates well with the degree of phase segregation measured by other techniques such as differential scanning calorimetry. Similar conclusions have been obtained by other investigators in toluene diisocyanate (TDI) based urethanes.[31-33]

Scope of the Present Work

In this work we are investigating how processing variables influence the segmental structure, molecular weight, and degree of
phase segregation of a linear polyether based polyurethane produced
by reaction injection molding. Differences in morphology and phase
segregation induced by different thermal histories will be qualitatively explained with the aid of a simplified ternary phase diagram.

In addition, some recent infrared evidence of phase separation
during hand mixed polymerization will be presented. The relationship between these results and those found in RIM samples will be
discussed.


EXPERIMENTAL

Materials

The polyurethane system used in this study consisted of a polyether soft segment (1256 Niax Polyol, Union Carbide Corporation)
with ca. 30% polyoxyethylene oxide as an end block on polyoxypropylene with a functionality of ca. 2.0 and $\overline{M}_n \simeq 2000$. The hard segment in the reaction injection molded samples is based on 4,4'-di-
phenylmethanediisocyanate, MDI (Mondur M, Mobay Chemical Company)
and 1,4-butanediol, BDO (GAF Chemicals Corporation). The infrared
samples were prepared using a modified form of MDI (Isonate 143L,
The Upjohn Company). This material is a liquid at room temperature,
with a functionality slightly greater than 2.0 and is chosen because
of easier handling at room temperature and sample characteristics
similar to those prepared from pure MDI.[8] The catalyst used was
dibutyltin dilaurate (DBTDL - T12, M & T Chemicals). Catalyst levels
in RIM samples are similar to those used in commercial processes.
The samples polymerized in the infrared study had catalyst levels
below commercial concentrations. Several studies have recently
been published using one or both polyurethane systems described
above.[7-10,18,42]

The polyol and BDO of the desired stoichiometry were mixed,
catalyst added, if required, and subsequently degassed at room temperature for several hours. The pure MDI was melted and filtered
at about 60°C immediately prior to each run while the liquid MDI
was used as received.

Sample Identification

Different samples were identified by the code given in Figure
1 which consisted of a letter to identify the type of isocyanate
(L = liquid MDI, M = pure MDI), the first two digits representing

the hard segment content (as total weight percent of isocyanate and extender), the following two or three giving a characteristic reaction temperature (wall temperature for RIM samples and polymerization temperature for isothermal IR samples) the last two digits providing the catalyst concentration. The set of all the samples used in this study is presented in Tables I and II.

Reaction Injection Molding

The molded samples were produced on an improved version of our laboratory size RIM machine.[35] With this machine we are able to duplicate the properties achieved on large production machines. Table III compares physical properties obtained from a Union Carbide commercial formulation for automotive fascia, using our machine, and those typical for production equipment.

The materials were injected in stoichiometric ratios into end-gated rectangular aluminum molds of 200 x 200 x 3 mm in dimension. Molds were preheated in an oven to the required temperature, and treated with a silicone mold release (Slipicone (R), Dow Corning). After injection, the plaques were demolded and placed in an oven for about one hour at the same temperature as the mold. After this, the samples were kept for at least five days at room temperature before further analysis. The initial monomer temperature was about 60°C in all cases.

Dynamic Mechanical Measurements

These measurements were performed in a Rheometrics Dynamical Spectrometer (RDS) using rectangular sample bars of 3 x 10 x 45 mm, at a frequency of 1 Hz and 0.1% strain over a temperature range from 173 K to 300 K.

SAMPLE IDENTIFICATION CODE

| M | – | 60 | – | 100 | – | 20 |
|---|---|----|---|-----|---|----|

| DIISOCYANATE TYPE: MDI | % HARD SEGMENT (WT% MDI + BDO) | REACTION TEMPERATURE (°C) | CATALYST CONCENTRATION ($G_{CAT}/G_{MIX} \times 10^5$) |
|---|---|---|---|

Figure 1. Sample identification code (see text).

Table I

Summary of RIM Samples

| Catalyzed | Uncatalyzed |
|-----------|-------------|
| M-60-60-20 | M-60-80-00 |
| M-60-100-20 | M-60-100-00 |
| M-60-105-20 | |
| M-60-64-75 | |
| M-60-100-75 | |
| M-60-134-75 | |

Table II

Summary of Infrared Samples

| Catalyzed | Uncatalyzed |
|-----------|-------------|
| L-30-27-10 | L-30-27-00 |
| L-50-27-10 | L-50-27-00 |
| L-60-27-10 | L-60-27-00 |
| L-70-27-10 | L-70-27-00 |

## Wide-angle X-ray Scattering

WAXS studies were carried out on 25 x 25 x 3 mm samples and
reflected intensity was recorded as a function of scattering angle
($2\theta$) at the rate of 1°/min, with a Siemens D500 diffractometer using
monochromatic Cu-K$\alpha$ radiation.

## Molecular Weight Characterization

Relative molecular weights and molecular weight distributions
of the RIM samples were determined using a duPont 830 liquid chro-
matograph, equipped with a duPont 833 flow controller. Three du-
Pont SEC silica columns with pore sizes of 60, 100 and 500 Å were
connected in series with a refractive index detector (LDC, model
1107). About 1 wt. % of polyoxyethylene ($\overline{M}_n \simeq 1000$) was added to
the solvent to minimize adsorption.

## Optical Microscopy

Samples were microtomed using an American Optical microtome.
Samples were frozen in liquid nitrogen and sectioned with a glass
knife. Sections were examined with a Zeiss polarizing microscope
using an 80 X magnification.

Table III

Comparison of Mini-RIM Plaque Mold Properties to
Typical Properties of RIM 2600 (Union Carbide)[41]

|  | Typical | Mini-RIM |
|---|---|---|
| **Process Conditions:** | | |
| Polyol Temperature (°C) | 55 | 60 |
| Isocyanate Temperature (°C) | 23 | 30 |
| Mold Temperature (°C) | 71 | 100 |
| Post-cure (hr./°C) | 1/121 | 1/100 |
| **Physical Properties:** | | |
| Hardness, Shore | 58 | 55 |
| 100% Modulus (Pa) | $1.6 \times 10^7$ | $1.8 \times 10^7$ |
| Tensile Strength (Pa) | $2.3 \times 10^7$ | $2.4 \times 10^7$ |
| Ultimate Elongation (%) | 235 | 200 |
| Flex Modulus (Pa): | | |
|     -29°C | $4.8 \times 10^5$ | $6.7 \times 10^5$ |
|     24°C | $1.8 \times 10^5$ | $1.9 \times 10^5$ |
|     70°C | $1.0 \times 10^5$ | $0.9 \times 10^5$ |

## Fourier Transform Infrared Spectroscopy (FTIR)

Infrared spectra were recorded using a Digilab FTS-10 inter-
ferometer, interfaced with a real time disc operated data system.
Data collection, Fourier transformation, file storage of spectra,
and subsequent information retrieval and display were performed using
the standard software supplied with the instrument.  The principle
of operation of FTIR[36,37] and its advantage in studying kinetics
of relatively fast systems[36,38] have been given elsewhere and will
not be discussed here.

A detailed description of the experimental procedure has recently been described by Camargo et al.[9]  Pre-measured amounts of
the monomers are mixed mechanically for less than a minute.  A small
sample, typically a drop, is placed between standard IR salt (NaCl)
plates and pressed to a thickness between 10 and 30 μm.  A sample
collection program is then activated.  It requests a series of inter-
ferograms at fixed time intervals.  These, along with a reference
spectrum (air) are stored, Fourier transformed, and an absorbance
spectrum obtained for each at the end of each run.

All samples were run at room temperature (27 ± 2°C) using 8
scans per spectra at 8 cm$^{-1}$ resolution.  This last condition gives
a time per scan of ca. 0.8 sec.

RESULTS

## Dynamic Mechanical Measurements

Dynamic shear moduli, G' and G", are shown in Figure 2 for a catalyzed and uncatalyzed sample of the same stoichiometry, both at a wall temperature of 100°C. The catalyzed sample shows a continuous drop in the elastic modulus for temperatures above the glass transition temperature of the soft segments, Tg. The Tg value, measured from the maximum in G", was found to be larger for all the catalyzed samples.

The behavior of the loss tangent is also significantly different for the catalyzed and uncatalyzed samples, as seen from Figure 3. The latter present a definite peak maximum near the soft segment Tg, while the catalyzed samples show a continuous broadening with no definite maximum after the soft segment Tg.

Results for different samples are summarized in Table IV. The ratio between the elastic modulus, G', at two temperatures (20°C and -30°C) gives a measure of the flatness of the rubbery plateau. Significant differences are observed between catalyzed and uncatalyzed samples. These can also be compared to a value of 0.35

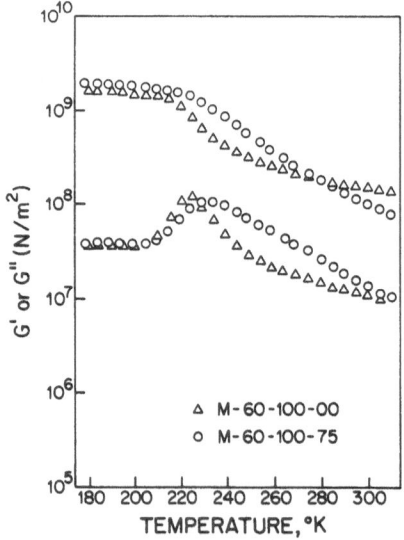

Figure 2.  Comparison of dynamic moduli, G' and G", for a catalyzed (M-60-100-75) and an uncatalyzed (M-60-100-00) sample.

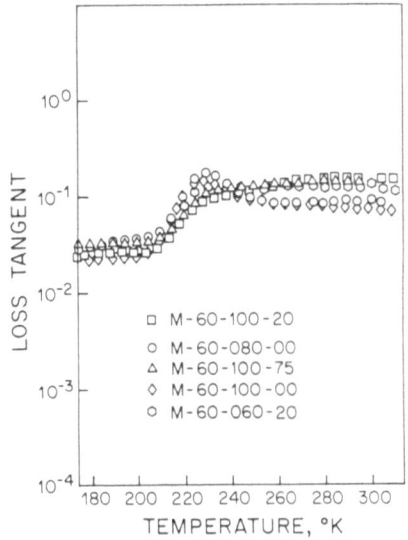

Figure 3.   Loss tangent for several catalyzed and uncatalyzed
            samples.  Catalyzed samples do not show a definite
            maximum peak over the temperature range studied.

Table IV

Summary of Dynamic Mechanical Results

|              | $G'(-193K) \times 10^{-9}$ $(N/m^2)$ | $G'(293K)/G'(243K)$ | $(Tg)s$ $(K)$ |
|--------------|--------------|--------------|--------------|
| M-60-80-00   | 1.10         | 0.45         | 222          |
| M-60-100-00  | 1.44         | 0.47         | 221          |
| M-60-60-20   | 1.65         | 0.23         | 224          |
| M-60-100-20  | 1.25         | 0.20         | 227          |
| M-60-105-20  | 1.55         | 0.21         | 226          |
| M-60-64-75   | 1.59         | 0.17         | 228          |
| M-60-100-75  | 1.82         | 0.23         | 224          |
| M-60-134-75  | 1.12         | 0.20         | 227          |

reported for pure MDI hand cast samples by Zdrahala et al[19] and a
value of ca. 0.10 for RIM catalyzed samples with 46% hard segment
based on liquid MDI.[43]   In summary, catalyzed RIM samples have a
less complete phase separation when measured by dynamic mechanical
behavior.

## Wide Angle X-ray Scattering

Figure 4 shows intensity vs. angle WAXS curves for several samples. Non-catalyzed samples have strong to medium crystalline reflections at 3.49, 3.77, 4.07, 4.57, 4.81 and 8.30 Å. These values are in good agreement with those reported by other investigators[6,7,23] (cf. Ref. 23, Figure 9). The catalyzed samples show very weak reflections indicating lower hard segment crystallinity.

Comparison of these results indicate better hard segment organization and domain purity in the uncatalyzed samples. WAXS results, however, have to be interpreted with care, since they only reach parts of the sample next to the skin. Thomas and co-workers[7] have calculated the depth of penetration for similar experimental conditions to be of the order of 100 μm.

## Optical Microscopy

Examination of the cross section by optical microscopy reveals strong differences between catalyzed and uncatalyzed samples (Figure 5). The uncatalyzed material is very birefringent under polarized light, but the micrograph does not give enough detail to confirm the presence of spherulites. The cracks observed give an indication of the brittle nature of these samples.

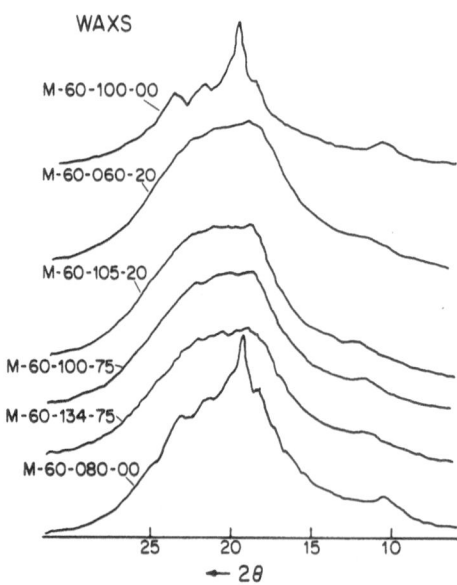

Figure 4.   Summary of WAXS results.  Samples identification according to Figure 1.

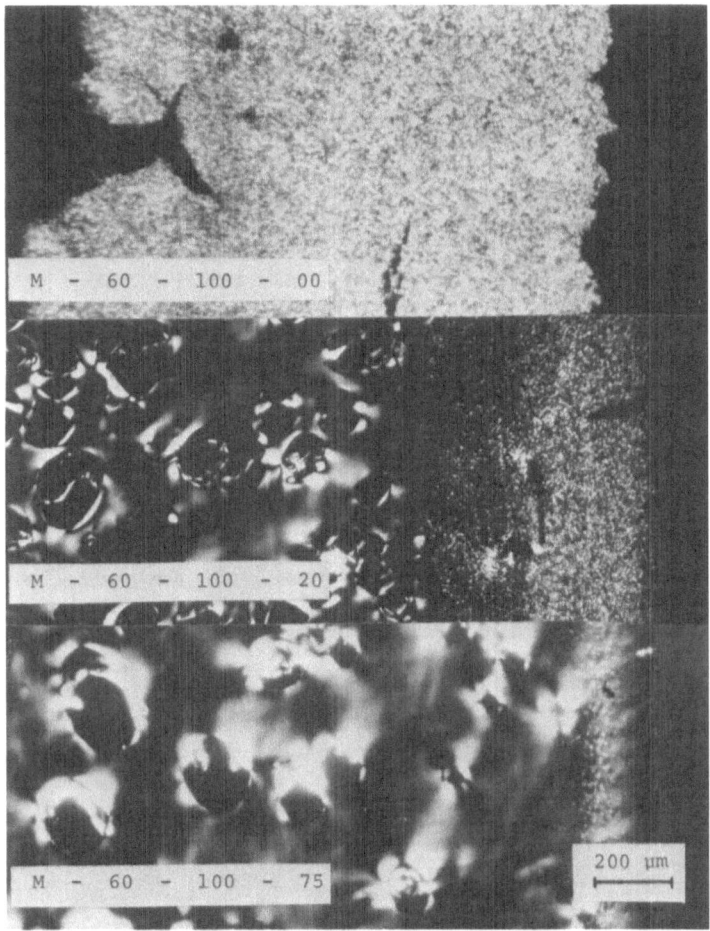

Figure 5.  Optical micrographs of samples containing 60% hard seg-
          ment, but with different catalyst.  The bubbles in the
          center of the two catalyzed samples are due to foaming
          during the cure and are not believed to affect the
          morphology.


    The core of the catalyzed samples generally appears homogeneous
although some bubbles are observed also.  The population of spherulites near the mold surface (skin) decreases with increasing catalyst concentration making the sample less susceptible to brittle
surface cracking.

Molecular Weight Determination

    Figure 6 shows GPC traces of a catalyzed (M-60-100-75) and un-
catalyzed (M-60-100-00) sample.  Lack of proper calibration curves

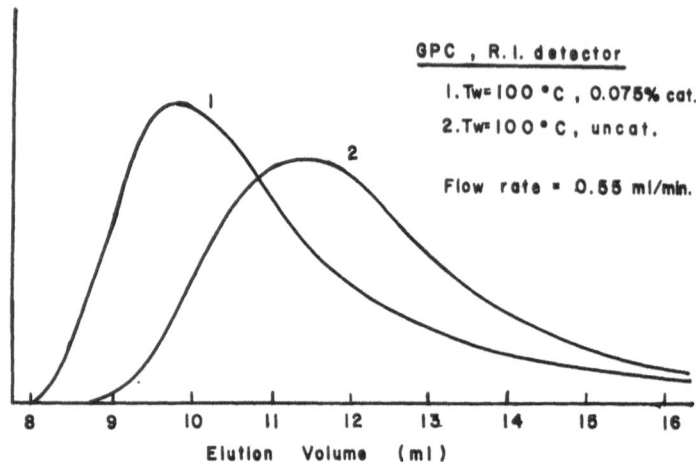

Figure 6.   Comparison of GPC traces for two 60% hard segment sam-
            ples, one uncatalyzed (M-60-100-00) and the other cata-
            lyzed (M-60-100-75).   Solvent:   dimethyl formamide.


precluded measuring absolute molecular weights.

    The relative trend in molecular weight, determined from peak
position on the elution volume axis, is clear:  catalyzed samples
have a higher molecular weight and narrower distribution than their
uncatalyzed counterparts.  This in part explains the brittle behavior
in the spherulitic uncatalyzed samples since the number of inter-
spherulite links would be expected to be low.

## Fourier Transform Infrared Spectroscopy

    Infrared spectra of a typical urethane polymerization at the
beginning and toward the end of the reaction are compared in Figure
7.  Several regions are of interest.  The strong absorption at ca.
2270 cm$^{-1}$ corresponds to the isocyanate asymmetric vibration and
can be used both to monitor conversion and thus reaction kinetics.[9,44]
The $CH_2$ symmetric vibration at ca. 2870 cm$^{-1}$ can be used as a reference band to compensate for any thickness changes in the sample.[9,44]

    The region between 3100 and 3600 cm$^{-1}$ is dominated by a broad
hydroxyl absorption during early stages of the reaction (Figure 7a),
but the hydrogen bonded amide hydrogen frequency ca. 3320 cm$^{-1}$ is
easily identified at larger reaction times (Figure 7b).  However,
the overlapping of absorption frequencies makes this region difficult
to analyze.

    The carbonyl region (1600-1800 cm$^{-1}$), which has little inter-
ference from other bands, is more suitable for H-bonding studies.
Figure 8 is a time sequence of the carbonyl absorption bands.  The

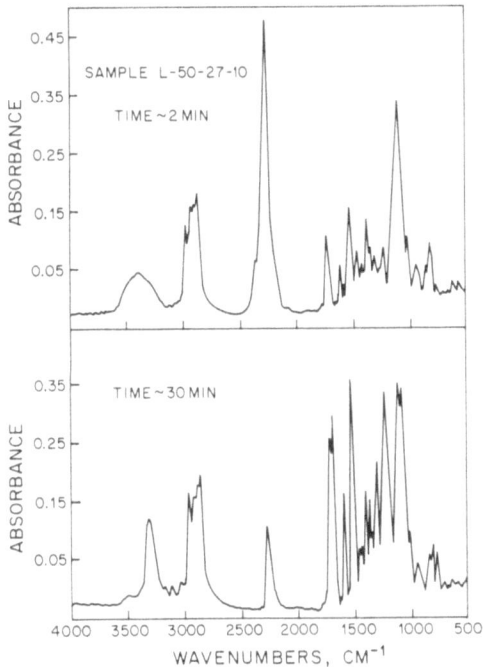

Figure 7.  Comparison of infrared spectra for a reacting polyurethane
system (sample L-50-27-10).  (a) at ca. 2 min.; (b) at
ca. 30 min.  Important for comparison is the region between 1600 and 3800 cm$^{-1}$ (see text).

free carbonyl band at 1730 cm$^{-1}$ dominates at short reaction times
but reaches a maximum at some intermediate conversion.  While the
absorption band of the bonded carbonyl groups at 1705 cm$^{-1}$ is low
for the first part of the reaction, an accelerated growth is observed at the point at which the free band reaches its maximum.

Strictly, the bands should be resolved and integrated absorbances used in the analysis.  If it is assumed that little or no
broadening takes place in a particular band during the reaction, and
that extinction coefficients for free and carbonyl bands are almost
equal,[45] integrated absorbances are roughly proportional to the
heights.  We have verified this for some of the samples.  A plot of
normalized peak heights versus conversion for one of the samples is
presented in Figure 9.  The same trend discussed before applies.
There is a narrow interval of conversion, in the range of 0.3 to
0.6, depending on the particular sample, in which hydrogen bonding
of carbonyl groups is accelerated as the hard segments associate,
presumably as phase separation ensues.[9]

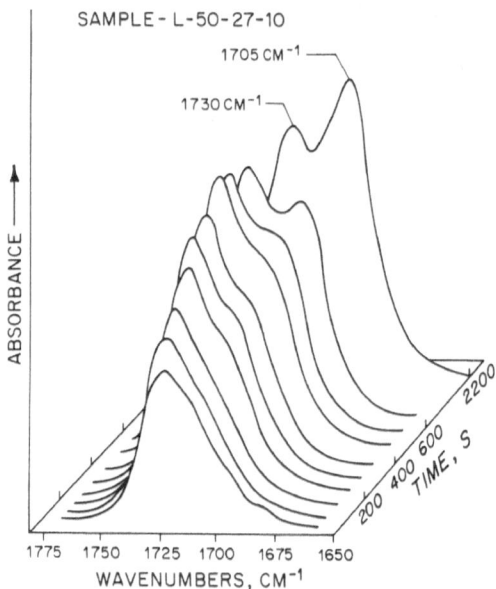

Figure 8.   Time sequence of the free (1730 cm$^{-1}$) and hydrogen bonded
(1705 cm$^{-1}$) carbonyl bands.   Sample L-50-27-10.


DISCUSSION

Reaction Injection Molded Samples

There are marked differences between RIM uncatalyzed samples
and those produced with different catalyst levels typical of RIM
processes.   Catalyzed materials are incompletely phase separated
solids reinforced by a disperse, probably highly interconnected,
hard phase which yields a tough flexible material.   Lower soft segment Tg, flatter elastic modulus rubbery plateau, higher hard segment, X-ray crystallinity, and the observation of large spherulites
all support the view that uncatalyzed samples are more completely
phase separated.   The low molecular weight found for these samples,
coupled with the localization of physical crosslinks in a few large
spherulites explains the apparent lack of significant interspheru-
lite connectivity and consequent low fracture resistance.

Thermal histories of the samples in the mold can be calculated
using a curing model for rectangular molds developed by Macosko and
co-workers.[39,40]   Figure 11 shows the results of this model for
centerline profiles, for a wall temperature of 100°C comparing the
three catalyst levels used in this study.   The two catalyzed samples display sharp temperature rises in the first few seconds that
attain the melting region of the hard segments.[7]   Conversion profiles, not shown here, indicate that the reaction is essentially

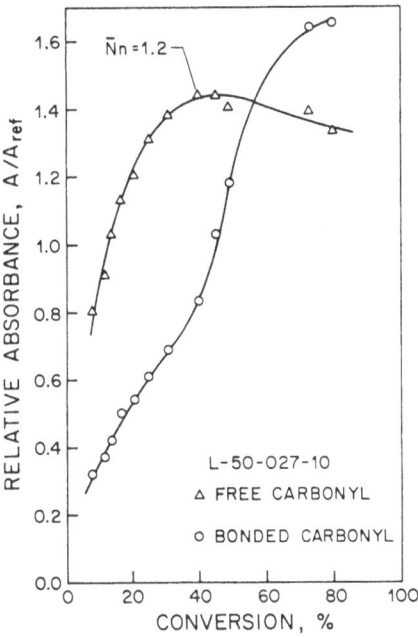

Figure 9.   Infrared absorption heights of the bonded and free car-
            bonyl peaks as function of isocyanate conversion.  The
            number average sequence length at maximum free carbonyl
            concentration is indicated.

complete at the point of maximum temperature.  Thus any morphology
that does form will be determined by a rate of supercooling controlled
only by heat conduction effects.

     Since the slower uncatalyzed reaction does not generate enough
heat to overcome heat conduction to the cold mold wall, the temperature in the center of the mold does not rise much above the mold wall
temperature.  The conversion is still low ( ~ 0.5) when the maximum
temperature is achieved and thus much of the morphology forms while
reactions take place at relatively low temperatures.  In fact, as we
will see later, reaction may be the controlling step in the domain
segregation-morphology formation process.

     Although the previous analysis is valid for the mold center line,
its qualitative features are also applicable to spatial variations
within a given sample.  Intuitively, as the mold wall is approached,
heat conduction becomes more important preventing the temperature at
these locations from rising as much as that of inner positions, consequently slowing the reaction even more.  The difference between
time required for completion of the reaction and morphology formation times decreases, resulting in morphological changes across the

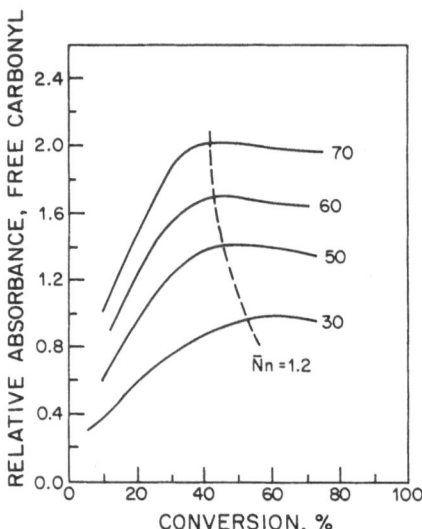

Figure 10.   Summary of free carbonyl absorption profiles as a func-
             tion of hard segment content.  Dashed line is the locus
             of conversions at which $\overline{N}_n$ = 1.2 for a given hard seg-
             ment concentration.


mold silimar to those that have already been observed by Thomas and
co-workers[7] in hand cast samples.

     Comparison of the optical micrographs of hand cast and RIM
molded material is quite revealing.  Reactions taking place at high
temperatures produce regions of little or no birefringence, but
significant DSC crystallinity[7] which exhibits a lower melting temperature than is observed in spherulitic regions.  Undoubtedly small
crystallite size and/or paracrystalline organization of hard segments are responsible for the lower melting point.  These feature-
less regions are considerably more ductile than the spherulitic regions which are prone to crack along interspherulitic boundaries.

     Partial segregation of the materials caused by poor impinge-
ment mixing produces large changes in morphology and molecular
weight on a more local 10-100 $\mu$ level.[41,42]  Location of the isocyanate and polyol fractions in different lamellar regions favors
the reaction between extender molecules and isocyanate to form
pure hard segment oligomers of sufficiently low solubility that they
precipitate into a separate phase (spherulitic or other) before they
have a chance to react with polyol molecules to produce multiblocks
of high molecular weight.[41,42]

     A simplified phase diagram (Figure 12) helps us to visualize
the phase separation in this system.  It is assumed in what follows

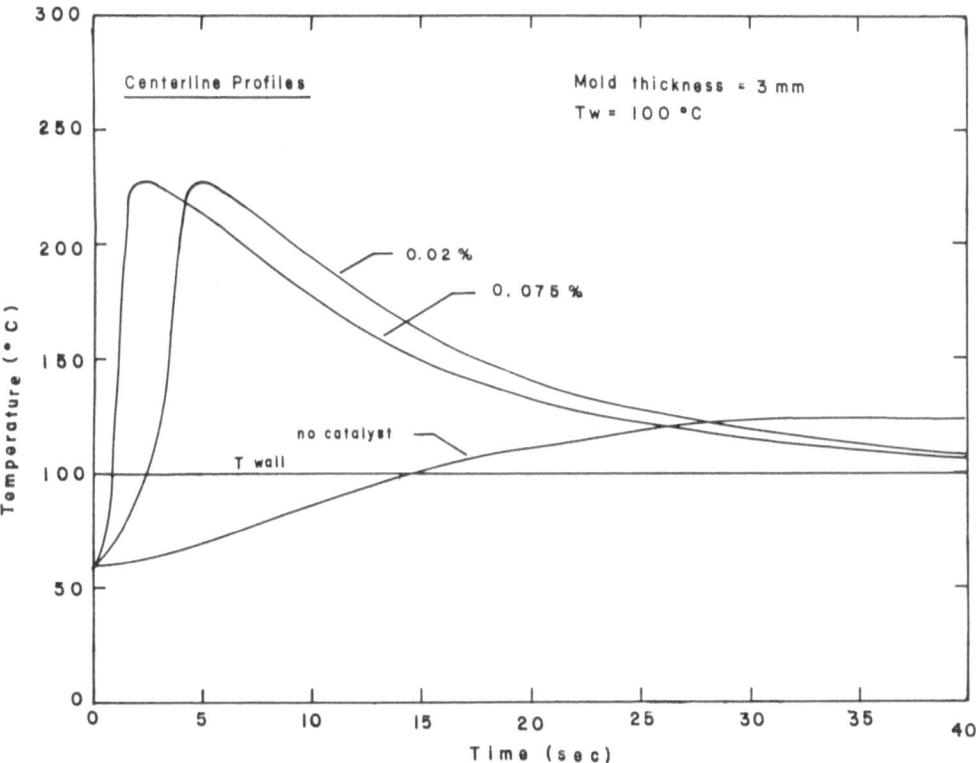

Figure 11.   Expected temperature profiles at the center of the mold.
These were obtained using a curing program developed by
Macosko and co-workers.[39,40]

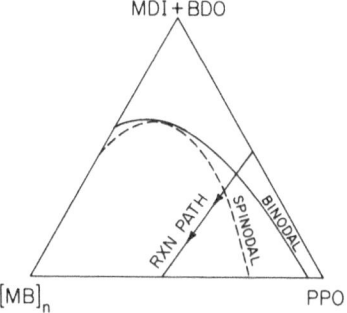

Figure 12.   A hypothetical phase diagram for the system polyol,
diisocyanate plus extender (MDI + BDO) and formed hard
segment, [MB]. From stoichiometry it can be shown that
reaction proceeds along the straight pathway shown.

that the initial reaction components exist in one homogeneous phase,
that the reaction proceeds isothermally, so that the binodal and
spinodal do not change with time and that there exists a local equilibrium at each point in the reaction.  As a rough approximation
of the true behavior, MDI and BDO are grouped together as one component since they should have similar solubilities in the polyol and
hard segment phase.  If the reaction is slow (e.g., uncatalyzed system), the mixture may spend enough time in the low supercooling region of the metastable region for nucleation and growth to occur at
relatively low primary nucleation density producing large spherulite
size.

The high reaction temperature produced in the catalyzed system
assures that reaction occurs in the one phase region and that high
molecular weight is produced.  As the material cools into the metastable region phase separation is resisted by the high viscosity of
the polymer and consequently nucleation and growth will occur only
at high supercooling where nucleation density is high.  Alternative-
ly the material might traverse the metastable region unchanged into
the spinodal region where mutually interpenetrating phases with mixed
phase boundaries will form.[46-50]  Solutions of polystyrene and 2,6-
dimethyl polyphenylene oxide of sufficiently high molecular weight
will spinodally decompose if they are cooled rapidly enough.[48]  In
both cases the sample will be made fracture resistant by an interconnected crystalline phase.

In real RIM samples the picture is more complex.  The boundaries defining the unstable and metastable regions may change dras-
tically with temperature.  More than one kind of phase separation
mechanism may take place at the same time within a sample due to
differences in thermal histories across the mold.  However, the
limiting situations discussed above should apply on a local basis.

## Fourier Transform Infrared Spectroscopy

As discussed above the development of the C=O absorption bands
in the 1600-1800 $cm^{-1}$ region can be studied, and an early detection
of phase separation is possible.  The maximum in the absorption of
the free C=O, accompanied by an increase in hydrogen bonded C=O at
an intermediate conversion can be identified with the onset of phase
separation during the polymerization.[9]

The number average sequence length of hard segments, $\bar{N}_n$, at the
maximum of the free C=O absorption peak was calculated using the
expression derived by Lopez-Serrano et al[51] and found to be in the
neighborhood of 1.2 for all samples studied independent of hard segment and/or catalyst concentration.  Figure 10 presents the free
carbonyl absorption profiles for several hard segment concentrations.
The dotted line was calculated from:[51]

$$p = \left[ \frac{1}{r_1} \left( 1 - \frac{1}{\overline{N}_n} \right) \right]^{1/2}$$

where p is the fractional conversion, $r_1$ is the initial ratio of BDO to MDI moles, $\overline{N}_n$ is the average sequence length of hard segments, taken as exactly 1.2 for all cases. The agreement between the maximum of each curve and the points of intersection is excellent, indicating that within experimental error, the maximum is achieved at a constant value of $\overline{N}_n$.

Infrared results are exactly the same as those calculated from turbidity data.[8] Clearly the scales of the two experiments are different, light transmission being sensitive to the formation of large crystalline structures such as spherulites while IR is sensitive to more local conformations and associations. We believe that the similarity in the calculated $\overline{N}_n$ values is due to a very fast spherulite growth subsequent to the hard segment association process that is detected by IR. Furthermore, it is possible that the discontinuity in DSC traces of the heat of reaction observed by Hager et al[10] in polymerization studies of similar urethane systems, and which also occurs at $\overline{N}_n \simeq 1.2-1.3$, be caused by the heat of crystallization accompanying the formation of spherulites during the sudden development of turbidity.

Infrared and light transmission experiments were carried out at low temperatures and catalyst concentrations where the formation of large spherulites is favored. However, no comparable experiments have yet been initiated to determine the onset and development of phase separation in fast reacting systems. Although we have observed that in general the opacity of RIM samples decreases as reaction speeds are increased (increasing mold wall temperatures, for example) mechanical measurements have shown that at least partial phase separation ultimately occurs.[52] Calorimetric studies of hand cast samples[7] have also shown that some crystallinity exists in clear, non-birefringent cores at high hard segment concentration.

An increased degree of phase mixing can generate a larger number of urethane group associations in a predominantly soft segment matrix. The validity of the IR analysis of phase separation has been briefly discussed in terms of hydrogen bonding on NH groups to the urethane alkoxyl oxygen.[31] If a significant amount of NH groups are bonded to the alkoxyl oxygen the fraction of NH which is hydrogen bonded within the hard phase is no longer given by the fraction of hydrogen bonded C=O groups. Other factors must, however, be considered. If hydrogen bonding in a given group is accomplished within an ordered structure, such as crystal or liquid crystal, larger frequency shifts may be possible. The occurrence

of urethane-urethane associations in the soft phase is likely to
have a more random organization than the one occurring inside crystalline hard segments.  Solution studies of model urethanes[31,34]
show that the C=O frequency shift in solution is smaller than that
reported for crystalline hard segments.[11,28-33]  Paik Sung and
Schneider[31,32] have also reported lower frequency shifts in 2,4-
toluene diisocyanate urethanes than in the crystalline 2,6-toluene
diisocyanate systems.  More recently, Senich and MacKnight[33] have
reported an extra shoulder which they did not identify and which
appears as a negative band when absorbance spectra are subtracted
from a reference spectrum at a lower temperature.  The frequency of
this shoulder is intermediate between those of the free and bonded
carbonyl groups (ca. 1725 $cm^{-1}$) and it is possible that it might
be due to the disruption of <u>crystalline</u> hydrogen bonding caused by
increasing temperature.

In summary, infrared experiments must be used with caution to
estimate phase separation in partially segregated polyuretnanes
such as those resulting from fast RIM polymerization.  Multiple carbonyl peaks[32] may result not only from hydrogen bonding in the crystalline and random state but also perhaps due to other effects such
as dipole-dipole coupling.[53,54]  Polymerizations carried out above
the melting temperature of the hard segments would provide valuable
information in this respect.  In this case any C=O hydrogen bonding
has to be random.  Specifically, an experiment can be designed to
compare the carbonyl absorption band for a sample slowly cooled down
from the melt (ca. 240°C) to that of a sample rapidly quenched.  In
the first case a gradual appearance of the bonded carbonyl band assigned to crystalline hard segments would take place.  In the
quenched sample the frequency shift may be lower due to the presence
of randomly hydrogen bonded C=O groups.  It is perhaps due to the
smaller frequency shift in the melt that MacKnight and Yang[29] could
only see a broad unresolved band at high temperatures in their urethanes.

No evidence of dipole-dipole coupling of carbonyl bands is
available and its occurrence is more difficult to identify.  Krim
and Abe[53] have carried out a perturbation analysis in amide I vi-
brations of crystalline polypeptides, and have shown that multiple
peaks may result from interactions between a hydrogen bonded C=O
group and the neighboring ones.  In polyurethanes it is likely that
only the interactions between C=O groups of adjacent urethanes is
important due to the chemical structure of the system, but the presence of non-crystalline interactions in partially phase separated
systems renders the analysis more complex.  We will use crystalline
hard segment model compounds to establish the number and positions
of the C=O peaks.  Perfectly crystalline material should show at
least two different bonded C=O peaks.[53]  Therefore, the idea of
bonded, non-bonded carbonyl bands must be carefully revised before

additional infrared studies of phase separation are undertaken.

## SUMMARY

The phase separation process largely determines the morphology which develops in RIM parts. Some of the methods used to follow phase separation, in particular dynamic mechanical spectroscopy, wide angle X-ray scattering and optical microscopy have been used to identify differences in phase separation developed during RIM urethane polymerizations with different rates of reaction.

It has been shown that slow reaction morphologies can be explained by spherulite growth that eventually limits phase connectivity and molecular weight development. Fast reaction on the other hand drives the reactive mixture into a high temperature one phase region, producing high molecular weight. Cooling after the reaction is essentially complete produces a fine interconnected microstructure.

Monitoring of slowly reacting samples at low temperatures by infrared spectroscopy shows that there exists a characteristic number average sequence length of hard segments at which phase segregation seems to begin. This is in agreement with previous studies by light transmission and differential scanning calorimetry in similar systems.

Based on the differences in phase separation between slow and fast systems, it is concluded that infrared spectroscopy and light transmission experiments in slowly reacting systems cannot be generalized to predict phase separation in fast RIM systems without further experimental work.

Clearly it is necessary to get a better understanding of the nature of premature phase separation in segmented polyurethanes that can limit molecular weight to less than optimum values. Viscosity rise, morphological development and gelation during RIM processing are determined by phase separation and all affect the mechanical properties of the part.

Phase separation studies are a major part of our current research in the area and we will continue studying the differences between slowly reacting and fast reacting systems. A detailed knowledge of the structures that can be formed can be used to modify processing variables to optimize the most desirable sample structure. More information is being gathered from infrared and light transmission studies in highly reactive systems. Other techniques such as small angle X-ray scattering are being considered for complementary studies of phase separation <u>during</u> the reaction.

## Acknowledgements

Financial support for this work was provided by the National Science Foundation, Polymers Program (Grant DMR79-09726) and by the Union Carbide Corporation, South Charleston, W. Va. A grant from the University of Minnesota Computer Center was provided for the calculations.

The authors also gratefully acknowledge the help of Mr. Robert Briber of the University of Massachusetts in the optical microscopy results and the help of Mr. Lou Moren in the RIM samples preparation. A fellowship from the 3M Company to R. E. C. is also acknowledged.

## References

1. J. M. Castro, V. M. Gonzalez and C. W. Macosko, Soc. Plast. Eng. Tech. Papers, 27, 363 (1981).
2. M. Tirrell, L. J. Lee and C. W. Macosko, A.C.S. Symp. Series, 104, 149 (1979).
3. G. D. Lewis, Paper #18, A.C.S. Rubber Div., 119th Meeting, Minneapolis, June, 1981.
4. L. J. Lee, Rubber Chem. and Tech., 53, 542 (1980).
5. A. Noshay and J. E. McGrath, "Block Copolymers: Overview and Critical Survey," Academic Press, NY, 1977, Ch. 7.
6. I. R. Fridman, E. L. Thomas, L. J. Lee and C. W. Macosko, Polymer, 21, 393 (1980).
7. A. L. Chang, R. M. Briber, E. L. Thomas, R. J. Zdrahala and F. E. Critchfield, Polymer, submitted for publication (1981).
8. J. M. Castro, F. Lopez-Serrano, R. E. Camargo, C. W. Macosko and M. Tirrell, J. Appl. Polym. Sci., 26, 2067 (1981).
9. R. E. Camargo, C. W. Macosko and M. Tirrell, Paper #17, A.C.S. Rubber Div., 119th Meeting, Minneapolis, June, 1981.
10. S. L. Hager, T. B. MacRury, R. M. Gerkin and F. E. Critchfield, Polymer Preprints, 21 (2), 298 (1980).
11. V. W. Srichatrapimuk and S. L. Cooper, J. Macromol. Sci., Phys., B15, 267 (1978).
12. S. L. Cooper and A. V. Tobolsky, J. Appl. Polym. Sci., 10, 1837 (1966).
13. T. Kajiyama and W. J. MacKnight, Macromolecules, 2, 254 (1969).
14. F. H. Huh and S. L. Cooper, Polym. Eng. Sci., 11, 369 (1971).
15. J. L. Illinger, N. S. Schneider and F. E. Karasz, Polym. Eng. Sci., 12, 25 (1972).
16. C. G. Seefried, J. V. Koleske and F. E. Critchfield, J. Appl. Polym. Sci., 19, 2493 (1975); ibid, p. 2503; ibid, p. 3185.
17. G. A. Senich and W. J. MacKnight, Adv. in Chem. Series, 176, 97 (1978).
18. R. J. Zdrahala, R. M. Gerkin, S. L. Hager and F. E. Critchfield, J. Appl. Polym. Sci., 24, 2041 (1979).

19. R. J. Zdrahala, S. L. Hager, R. M. Gerkin and F. E. Critchfield, J. Elastom. and Plast., 12, 225 (1980).

20. R. Bonart, J. Macromol. Sci., Phys., B2, 115 (1968).

21. R. Bonart, L. Morbitzer and G. Hentze, J. Macromol. Sci., Phys., B3, 337 (1969).

22. R. Bonart, L. Morbitzer and E. H. Muller, J. Macromol. Sci., Phys., B9, 447 (1974).

23. N. S. Schneider, C. R. Desper, J. R. Illinger and A. O. King, J. Macromol. Sci., Phys., B11, 527 (1975).

24. J. Blackwell and K. H. Gardner, Polymer, 20, 13 (1979).

25. J. Blackwell, M. R. Najarajan and T. Hoitnik, Polymer Preprints, 21 (2), 303 (1980).

26. Y. M. Boyarchuk, L. Y. Rapport, V. N. Nikitin and N. P. Apukhtina, Polym. Sci., USSR, 7, 859 (1965).

27. T. Tanaka, T. Yokoyama and Y. Yamaguchi, J. Polym. Sci., A-1, 6, 2137 (1968).

28. R. W. Seymour, G. M. Estes and S. L. Cooper, Macromolecules, 3, 579 (1970).

29. W. J. MacKnight and M. Yang, J. Polym. Sci., C42, 817 (1973).

30. J. C. West and S. L. Cooper, J. Polym. Sci., C60, 127 (1977).

31. C. S. Paik Sung and N. S. Schneider, Macromolecules, 8, 68 (1975).

32. C. S. Paik Sung and N. S. Schneider, Macromolecules, 10, 452 (1977).

33. G. A. Senich and W. J. MacKnight, Macromolecules, 13, 106 (1980).

34. T. Yokoyama, Adv. in Urethane Sci. and Tech., 6, 1 (1978).

35. L. J. Lee and C. W. Macosko, Soc. Plast. Eng. Tech. Papers, 24, 151 (1978); U. S. Patent 4,189,070 (1979).

36. (a) P. R. Griffiths, "Chemical Infrared Fourier Transform Spectroscopy," Wiley Interscience, New York, 1975; (b) P. R. Griffiths, C. T. Foskett and R. Curbelo, Appl. Spectrosc. Rev., 6, 31 (1972).

37. J. L. Koenig, Appl. Spectrosc., 29, 293 (1975).

38. J. O. Lephardt and G. Vilcins, Appl. Spectrosc., 29, 221 (1975); H. W. Siesler, Polymer Preprints, 21 (1), 163 (1980).

39. L. J. Lee and C. W. Macosko, Int. J. Heat. Mass. Transfer, 23, 1479 (1980); ibid, Soc. Plast. Eng. Tech. Papers, 24, 155 (1978).

40. J. M. Castro and C. W. Macosko, A.I.Ch.I. J., accepted for publication (1981).

41. P. Kolodziej, M.S. Thesis, Dept. of Chemical Engineering and Materials Science, University of Minnesota, 1980.

42. P. Kolodziej, C. W. Macosko and W. E. Ranz, Polym. Eng. Sci., submitted (1981).

43. J. M. Castro and C. W. Macosko, unpublished results.

44. E. G. Richter and C. W. Macosko, Polym. Eng. Sci., 18, 1012 (1978).

45. G. C. Pimentel and A. L. McClellan, "The Hydrogen Bond," W. H. Freeman and Co., San Francisco, 1960.

46.   O. Olabisi, L. M. Robeson and M. T. Shaw, "Polymer-Polymer Miscibility," Academic Press, New York, 1979.

47.   J. W. Cahn, Trans. Metall. Soc., AIME, $\underline{242}$, 166 (1968).

48.   S. Wellinghoff, J. Shaw and E. Baer, Macromolecules, $\underline{12}$, 932 (1979).

49.   J. Gilmer, N. Goldstein and R. S. Stein, Am. Phys. Soc. Bulletin, $\underline{25}$, (3), 353 (1980).

50.   T. Nishi, T. T. Wang and T. K. Kwei, Macromolecules, $\underline{8}$, 227 (1975).

51.   F. Lopez-Serrano, J. M. Castro, C. W. Macosko and M. Tirrell, Polymer, $\underline{21}$, 263 (1980).

52.   R. E. Camargo, unpublished data (1981).

53.   S. Krimm and Y. Abe, Proc. Nat. Acad. Sci. USA, $\underline{69}$, 2788 (1972).

54.   J. Vincent-Geisse, Spectrochim. Acta, $\underline{24A}$, 1 (1968).

# GRAFT POLYOL RIM SYSTEMS: THERMAL STABILITY AND MOISTURE ABSORPTION

Robert A. Markovs

BASF Wyandotte Corporation

Wyandotte, MI 48192

## INTRODUCTION

Faced with legislative mandates requiring strict adherence to Corporate Average Fuel Economy (CAFE) standards, the automotive industry has been searching for means of improving the fuel efficiency of their fleet. Methods such as improving engine efficiency and design changes to give the best aerodynamic air flow have already been implemented. Another widely used approach is that of downsizing. As cars are made smaller, body weights decrease, improving fuel efficiency. However, to maintain adequate passenger comfort requirements in the interior of the vehicle, downsizing can only be carried to a practical limit.

To achieve further weight reductions of some 1000-1500 lbs., auto makers are considering tradeoffs to lower weight materials. Exterior automotive body panels are one such area where replacement of the traditionally stamped steel parts by plastic are being seriously considered. For example, a typical steel fender will weight 13 lbs. whereas a comparable plastic fender will weight only 6 lbs. Likewise, when one considers energy usage over the life of the part, including energy required to form the part, plastics once again emerge as clear winners. Potential automotive applications for plastic replacement parts include fenders, deck lids and door panels. Plastics for use in such applications must meet demanding requirements before their application can be considered. These include:

1. Sufficiently rigid to be self supporting.

2. Thermal dimensional stability to allow normal processing operations at elevated temperatures.

3.  Low coefficient of thermal expansion (CTE).

4.  Class A surface and good paintability.

5.  Good impact characteristics at low temperature.

These requirements can be largely met by Reaction Injection Molded (RIM) polyurethanes. The rigidity requirement can be met in various ways. The part may be designed with ribs or other structural
modifications so as to increase part stiffness. A higher modulus RIM
system can be employed. Milled glass or other fillers can be added
to the raw materials.

Low CTE is not inherent in polyurethanes. To achieve CTE's approaching that of aluminum (27 x $10^{-6}$ in/in, °F), milled glass or
other fillers must be introduced into the polymer matrix. In an exterior body panel application this is a very key property which can
be met by addition of fillers.

Class A surface and good paintability are properties inherent to
RIM polyurethanes. Even the addition of various fillers presents no
problem toward achieving this requirement.

Impact properties of RIM urethanes are also very good. While addition of fillers will give somewhat poorer impact properties, they
will still be good.

This leaves only Thermal Dimensional Stability as a key requirement requiring development work. The following discussion will ad-
dress itself primarily to RIM formulations having good thermal dimensional stability.


EARLY DEVELOPMENT

All major components comprising a RIM system were investigated
with regard to thermal stability:

1.  Chain extenders.

2.  Polyol.

3.  Isocyanate.

The most widely used chain extenders in RIM systems are butanediol and ethylene glycol. Both extenders were evaluated in terms of
their thermal stability. It was found that ethylene glycol is superior to butanediol in terms of thermal stability. When using the
heat sag test detailed later, systems containing butanediol consis-
tently had much greater sag than samples prepared using ethylene gly-

col containing systems.

Several polyols, both graft and non-graft, were investigated. Graft polyols are derived by free radical addition of styrene/acrylonitrile to a conventionally prepared polyol. Thermal properties of both conventional and graft polyols were found to be generally comparable. Heat sag testing of both types of systems showed that sags were very similar. However, graft polyol systems were found to process much better with the ethylene glycol extender required for maximal heat stability. Flow characteristics using a graft polyol system were improved in most cases. In addition, since graft polyols give higher modulus values than conventional polyols, it was possible to use less chain extender to achieve a given modulus. This is important because as chain extender levels are increased, green strength generally becomes poorer. As a result, graft polyol systems were selected as being superior in performance for this study.

Various types of isocyanates were also investigated. Crude, higher functionality isocyanates, such as Upjohn's ISONATE 191; urethane modified isocyanates, such as Upjohn's ISONATE 181; carbodiimide modified isocyanates, such as Upjohn's ISONATE 143L and a specialty urethane modified isocyanate produced by BASF Wyandotte, Number 46 Isocyanate. It was found that both the crude isocyanate and strictly urethane modified isocyanate gave poor thermal stability results. Both ISONATE 143L and Number 46 Isocyanate had good thermal properties. While ISONATE 143L was slightly better with regard to thermal properties such as heat sag and modulus ratio, Number 46 Isocyanate gave better elongation and impact properties. As a result both isocyanates were investigated during the development period.

Investigative emphasis, then, was placed on graft polyol systems extended with ethylene glycol using appropriate catalysts, and either ISONATE 143L or Number 46 Isocyanate on the isocyanate side of the system.


MATERIALS AND PROCESSING EQUIPMENT

For the resin portion of the system, PLURACOL Polyol 581, a 25 hydroxyl styrene/acrylonitrile graft triol, was used as the polyether.

For the isocyanate side of the system, two different isocyanates were evaluated. ISONATE 143L, a nominal 29.1% free NCO carbodiimide modified diphenyl methane diisocyanate (MDI) from Upjohn, and Number 46 Isocyanate, a nominal 26.0% free NCO urethane modified MDI from BASF Wyandotte Corporation.

Milled glass fiber, 1/16" Type B, P 117 from Owens Corning, was used as the filler for glass filler samples.

Unfilled plaques, 36 x 36 inches, were made using both a Krauss-Maffei PU 80/155 and an Elastogran Maschinenbau (EMB) PUROMAT 300 SV RIM machine. Glass filled plaques were made solely on the PUROMAT 300 SV. Machine outputs for the Krauss-Maffei machine ranged from 3.5–4.0 lb./sec. while PUROMAT 300 SV outputs ranged from 3–10 lb./sec.

SAMPLE CONDITIONING AND TEST METHODS

Standard test procedures were carried out after all samples had been conditioned for 24 hours at constant temperature (72°F) and humidity (50%).

The procedure for the crucial heat sag test is based on ASTM Method D3769. This method employs a 6 x 1 x 1/8" test specimen. The test specimen is clamped on one end only leaving a 4 inch unsupported overhang. The sample is placed in a forced draft oven at 125°C for 60 minutes. At the end of this period the vertical deflection from the original horizontal plane of this sample (in inches) is measured. This value is the heat sag. The above method produced very little differentiation of materials prepared in this investigation. Consequently, a modified version of the test using a test specimen 8 x 1 x 1/8 inch was used with a 6 inch overhang. Such a test specimen was placed in a forced draft oven at 325°F for 30 minutes. The differ - ence in the vertical height from the sample to the base of the clamp was measured before and after exposure. All samples were postcured under the indicated conditions within one hour of demold.

DISCUSSION

Thermally Stable RIM Systems

Heating of polyurethanes at various temperatures has been shown to have a definite effect on physical properties. Cooper[1,2] has shown that thermoplastic polyurethanes undergo complex ordering processes when heated at elevated temperatures. The process of heating a polyurethane material at elevated temperatures has been variously referred to as curing, post-curing or annealing. In this discussion the terms curing and annealing will be used interchangeably. Other workers have shown that microcellular polyurethanes produced by the RIM process undergo similar ordering processes.[4] These processes have been categorized as phase separation, molecular alignment and crystallization.

Curing of microcellular samples at temperatures ranging from 250°F to 400°F has been shown to have a definite effect on physical properties, especially heat sag. Figure 1 shows a plot of heat sag vs. annealing time at 325°F for the graft polyol/ethylene glycol sys-

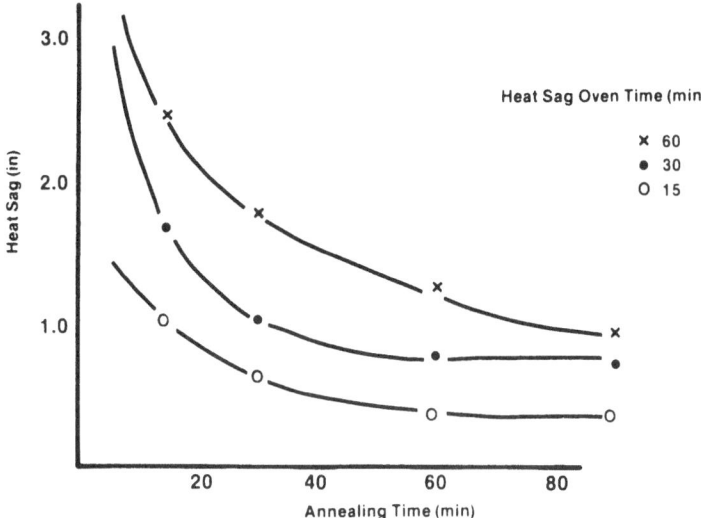

Figure 1. Annealing time vs. heat sag

tem using Number 46 Isocyanate, a urethane modified isocyanate.

The family of curves in Figure 1 is generated by measuring the sag of the test specimen after various time periods in the oven. For the standard 30 minute time period, the 325°F heat sag levels off at about 0.8 inch regardless of annealing time once an initial 30 minute annealing period has been reached.

By comparison, the same resin system with the carbodiimide modified isocyanate (ISONATE 143L) shows a slightly better annealing time/heat sag response (Figure 2). While a similar curve is generated, the carbodiimide modified isocyanate gives a 30 minute heat sag of about 0.6 inch.

Table I gives physical properties for the system using the carbodiimide modified isocyanate (WUC-23330R-143L) as a function of post cure temperature and time. As is evident, most physical properties are optimized by a 250°F post cure. Heat sag is the only property that requires a 325°F annealing step in order to achieve the best result. A similar phenomenon was noted with the use of Number 46 Isocyanate and the same resin system.

The WUC-23330R-143L RIM system was selected based on a combination of automotive requirements including processing parameters and physical properties. Work was conducted with various extender levels to arrive at this compromise with both the urethane modified and carbodiimide modified isocyanates. Figure 3 shows the effect of an increase in equivalents of ethylene glycol on room temperature modulus

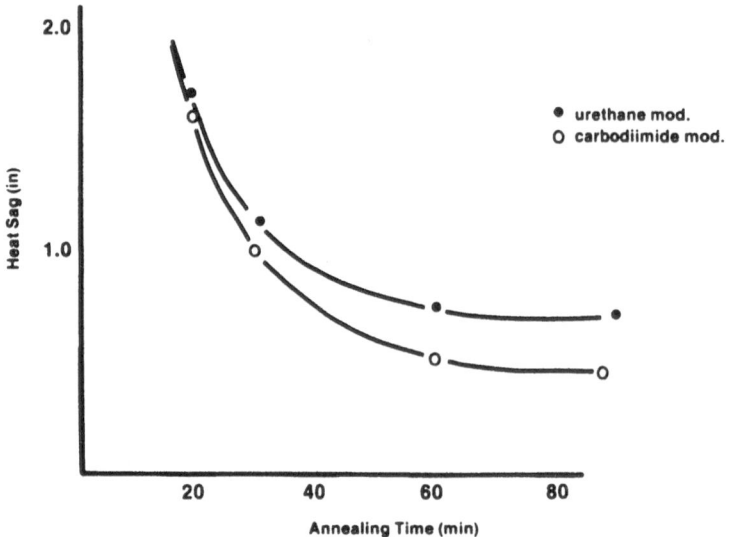

Figure 2.   Annealing time vs. heat sag

Table I

High Temperature Stable RIM System WUC-23330R-143L

| | | | |
|---|---|---|---|
| Post Cure Temperature, °F. | NO | 250 | 325 | 325 |
| Post Cure Time, minutes | NO | 60 | 30 | 60 |
| Density, pcf. | 68.0 | 67.5 | 69.7 | 67.9 |
| Tensile, psi. | 4175 | 4810 | 4755 | 4668 |
| Elongation, % | 62 | 155 | 150 | 165 |
| Graves Tear, pi. | 839 | 927 | 885 | 891 |
| Shore "D" Hardness | 59/59 | 60/60 | 60/60 | 60/60 |
| 250°F. Heat Sag, inches | - | 0.20 | 0.10 | 0.10 |
| Flexibility Modulus, K psi. | | | | |
| -20°F. | 200.1 | 171.6 | 164.1 | 169.7 |
| 72°F. | 89.8 | 87.7 | 86.5 | 89.8 |
| 158°F. | 47.9 | 49.8 | 54.2 | 55.1 |
| Ratio -20°F./158°F. | 4.19 | 3.45 | 3.04 | 3.08 |
| Notched IZOD Impact ft-lb./in. | 5.6 | 11.1 | 10.3 | 10.3 |
| 325°F. Heat Sag, inches | >3 | >3 | 1.44 | 0.80 |
| Coefficient of Thermal Expansion in./in., °F. X $10^{-6}$ | - | 75 | 72 | 70 |

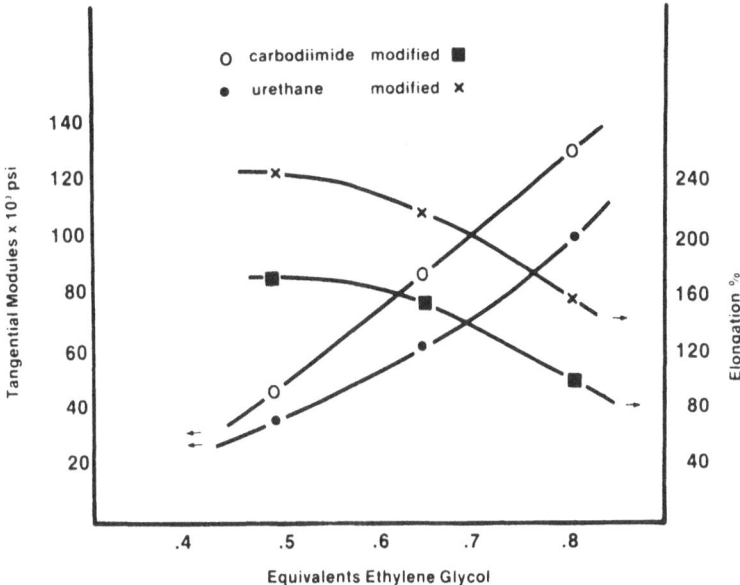

Figure 3.  Modulus and elongation vs. equivalents EG

and elongation with both isocyanates.  For the carbodiimide modified
system, modulus increase is fairly linear rising from 42,000 psi with
about 0.5 equivalents of ethylene glycol to 130,000 psi with about
0.8 equivalents of ethylene glycol.  For this same chain extender increase, elongation drops from 180% to 100%.  In a like manner, with
the urethane modified isocyanate, modulus increases from 30,000 psi
to 105,000 psi while elongation falls from 220% to 160%.

Because of stiffness and coefficient of thermal expansion requirements, the current state of the art uses milled glass fiber as
a filler for the polyurethane matrix.  The effect of addition of
milled glass fiber on the physical properties of the preferential
WUC-23330R-143L system was investigated.  Figures 4 and 5 show the
effect of annealing time vs. 325°F heat sag at glass loadings of 15
and 25%.  A family of curves is generated in both cases with respect
to the length of time the test specimen is in the oven for the heat
sag test.  With 15% glass the sample needs 20-30 minutes annealing
at 325°F to reach the optimal heat sag.  With 25% glass loading the
family of curves generated is similar.  However, the actual distance
of sag as expressed on the ordinate axis is much less.  To reach a
heat sag of 0.3 inches requires that the sample be annealed at 325°F
for 20-30 minutes.

Table II gives the physical properties for the 15% and 25% glass
loaded systems.  Determination of physical properties parallel or

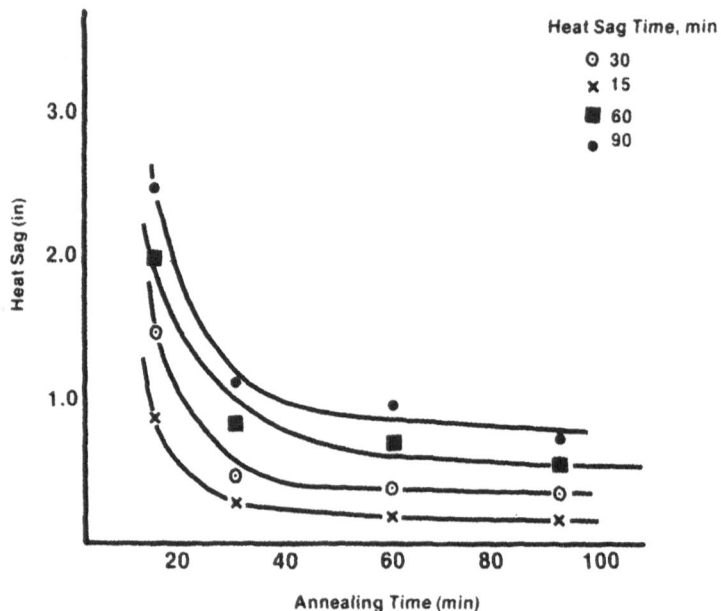

Figure 4.   WUC-2333OR-143L with 15% milled glass

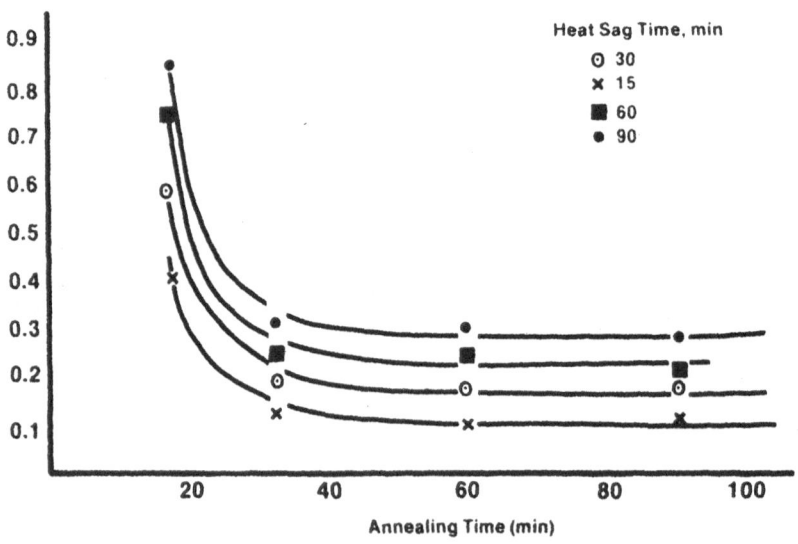

Figure 5.   WUC-2333OR-143L with 25% milled glass

Table II

Glass Filled WUC-23330R-143L

| Milled Glass, % | 15 | 15 | 25 | 25 |
|---|---|---|---|---|
| Flow Direction | Parallel | Perpendicular | Parallel | Perpendicular |
| Post Cure Temperature, °F. | 325 | 325 | 325 | 325 |
| Post Cure Time, minutes | 30 | 30 | 30 | 30 |
| Density, pcf. | 78.6 | 78.6 | 82.2 | 82.2 |
| Tensile, psi. | 5780 | 5750 | 6730 | 5940 |
| Elongation, % | 50 | 50 | 20 | 25 |
| Graves Tear, pi. | 880 | 846 | 953 | 890 |
| Shore "D" Hardness | 70/69 | 70/69 | 73/73 | 73/73 |
| 250°F. Heat Sag, inches | 0.10 | 0.10 | 0.03 | 0.06 |
| Flexibility Modulus X $10^3$ | | | | |
| $\quad$-20°F. | 258.3 | 236.9 | 382.7 | 287.3 |
| $\quad$72°F. | 217.1 | 175.9 | 302.7 | 207.8 |
| $\quad$158°F. | 132.1 | 99.8 | 166.0 | 121.4 |
| Ratio -20°F./158°F. | 1.95 | 2.37 | 2.31 | 2.37 |
| Notched IZOD Impact ft.-lb./in. | 7.30 | 6.00 | 6.10 | 5.70 |
| Coefficient of Thermal Expansion | | | | |
| $\quad$in./in., °F., $10^{-6}$ | 40 | 67 | 38 | 60 |
| 325°F. Heat Sag, inches | 0.34 | 0.67 | 0.15 | 0.53 |

perpendicular to the direction of flow shows the expected major differences with specific properties. Physical properties determined on a sample cut parallel to the direction of flow show better properties than one cut perpendicular to the direction of flow. For a 25% glass loading, room temperature modulus for samples cut parallel to flow direction is about 300,000 psi, while for the perpendicular sample, modulus is about 200,000 psi. Similarly, heat sag, tear, coefficient of thermal expansion and impact properties are all better when measured parallel to the flow direction. A greater degree of scattering of data has been found with glass filled systems than with unfilled systems. For 25% glass loaded systems, modulus values have been found to range from 250,000 to 350,000 psi.

## Moisture Absorption of Microcellular Foam

During the early portion of this work, it was noted that the foam test specimens that were annealed at above 250°F were not dimensionally stable. The length and width of the sample as well as the thickness increased. This swelling is thought to be caused by water trapped in the polyurethane matrix.

Initial studies on these systems was done using ethylene glycol having a high water content (0.11-0.14%). During the course of the

polyurethane formation reaction, apparently not all the water reacts
with isocyanate leaving water entrapped in the polyurethane matrix.
When annealed at high temperatures, one can hypothesize that the
water will be expelled from the polymer matrix by one of two ways or
a combination of both.  If unreacted isocyanate moieties remain in
the polyurethane, at elevated temperatures they could react with
water yielding carbon dioxide which would expand the foam while escap-
ing from the polymer matrix.  In a similar vein, the entrapped moisture could be forced from the polymer matrix by simple conversion to
water vapor which again would escape violently causing expansion of
the foam.  It is currently thought that the latter route is responsible for the swelling that is observed.

Ethylene glycol having a high and low water content was evaluated with both the carbodiimide modified and urethane modified isocyanates.  In one case ethylene glycol with 0.14% water was used
while in the other case ethylene glycol with 0.04% water content was
used.  Length, width and thickness of the test specimen was noted before and after annealing.  In all cases, the high water system gave
the most swelling.  Figure 6 shows that samples annealed at 350°F
expanded in thickness by almost 12% with the carbodiimide modified
isocyanate and 4% with the urethane modified isocyanate.  With the
lower water version, swelling remained at 1% or less.  In all cases
the foam expansion was greater with the carbodiimide modified isocyanate compared to the urethane modified isocyanate.

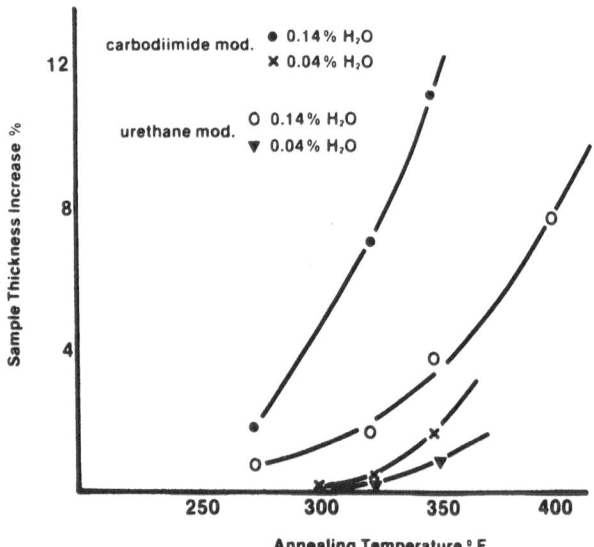

Figure 6.  Foam expansion vs. temperature

Foam expansion was also shown to be a function of time as well as temperature. Figure 7 shows that virtually all of the swelling occurs within the first 20 minutes of the heating process. After this period, no increase in swelling is noted.

If the test specimen is initially heated for a given time period at a lower temperature (300°F), then subjected to a higher temperature (350°F), the swelling will be less than if the sample were heated initially at the higher temperature alone. The mechanism here is believed to be one related to the violence of gas expulsion. When the foam sample is heated at the lower temperature, some of the entrapped moisture can diffuse out slowly without greatly expanding the foam. If the water in the polyurethane matrix is heated rapidly to a high temperature, the gas will expand rapidly and will be expelled more violently causing more swelling of the foam.

It was also demonstrated that the same swelling phenomenon can be made to occur in foam made using a low water system. Sample plaques were made using the low water system with the carbodiimide modified isocyanate, left unpost-cured, and conditioned in four different environments:

1. Dessicator

2. Ambient conditions (laboratory)

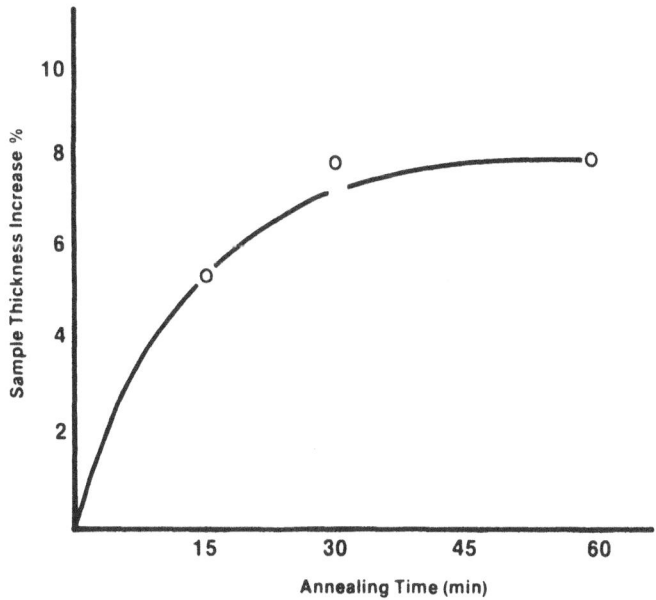

Figure 7.   Foam expansion vs. time (325°F)

3.  Constant humidity/temperature (50%/72°F)

4.  100% Humidity

Samples were maintained for various exposure times, then annealed
at 325°F.  Figure 8 shows the increase in sample thickness with respect to environmental exposure time.  In the case of the 100% humidity sample, uptake of moisture, as shown by the increased swelling, is rapid.  On the other extreme, the dessicated sample shows no
increase in swelling after a prolonged period.

Once the material has been annealed, moisture has little or no
effect on the sample.  An annealed sample, placed in the 100% humidity environment for one week, then annealed again at 325°F for one
hour, showed no swelling.

SUMMARY

A microcellular RIM system has been developed which gives excellent thermal stability at elevated temperatures.  Curing of the
microcellular foam at 250°F optimizes most properties, although to
achieve the best heat sag properties, the sample must be annealed at
325°F.

Addition of up to 25% milled glass to this system improves coefficient of thermal expansion and heat sag and more than triples the

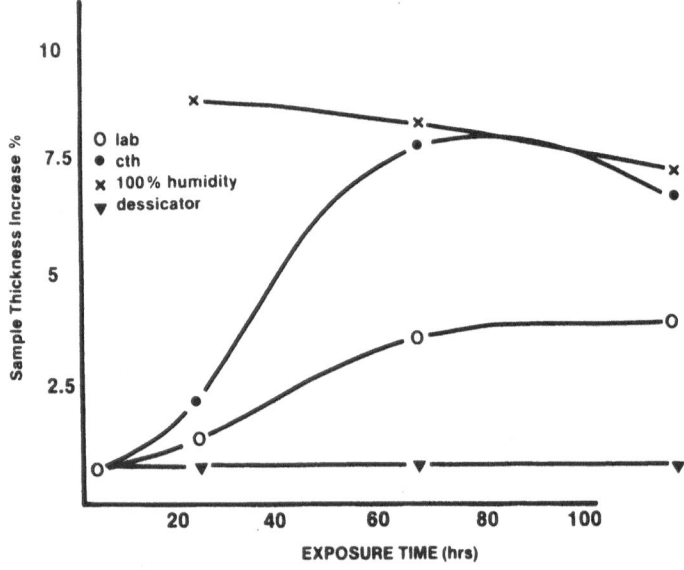

Figure 8.  Water absorption of microcellular foam

modulus.  As with the unfilled material, most physical properties are optimized by a 250°F postcure; a 325°F annealing step is needed to optimize heat sags.

It has been shown that water entrapped in the polyurethane matrix can lead to an undesirable swelling of the foam.  Water can be introduced into the polymer matrix by one of two ways.  The water may be present in the resin side before the polymerization reaction, or water may be absorbed by the uncured foam if allowed to remain in high humidity environments.  Ideally, the resin side of the RIM system should contain less than 0.04% water and the resultant polyurethane should be cured as soon as possible after preparation.

## References

1. S. L. Cooper and A. V. Tobolsky, J. Appl. Polym. Sci., 10, 1837 (1966).
2. H. N. Ng, A. E. Allegrezza, R. W. Seymour and S. L. Cooper, Polymer, 14, 255 (1973).
3. R. J. Zdrahala, R. M. Gerkin, S. L. Hager and F. E. Critchfield, J. Appl. Polym. Sci., 24, 2041 (1979).
4. D. M. Rice and R. J. G. Dominguez, Polym. Eng. Sci., 20, 1192 (1980).

# POLYAMIDE RIM SYSTEMS

L. M. Alberino and D. F. Regelman

The Upjohn Company

North Haven CT

## CHEMISTRY

### Introduction

Reaction injection molding (RIM) processes utilize rapid chemical reactions allowing the production of solid parts from liquid monomers with reaction times on the order of seconds to minutes. Most isocyanate RIM reactions involve step growth polymerization via addition of active hydrogen compounds across the C–N isocyanate double bond.

$$R - X - H + \phi - N = C = O \longrightarrow \phi - NH - CO - X - R$$

$$X = O, NH$$

Products = urethane, urea, allophanate, biuret

Commercially available systems generally utilize low molecular weight diols, higher molecular weight polyols, diamines or usually mixtures of two or three of these active hydrogen compounds. Formulations are adjusted to give a range of products from elastomers to more rigid structural parts. Products generally contain mixtures of urethane, urea, allophanate and/or biuret functionalities.

Modifications of isocyanate RIM processes involving trimerization of isocyanates to introduce isocyanurate structures are used to give enhanced thermal stability to the parts produced.[1]

Work is underway to evaluate potential RIM processes that do not involve urethane or urea products. Epoxy, polyester, polyacry-

late and polystyrene systems all show required reactivity without the
evaluation of volatiles.[2,3]  Additionally, the anionic ring-opening
of caprolactam has been investigated as an avenue for potential polyamide RIM systems.[3]

    Our attention has recently focused on polyamide RIM systems using
isocyanates as precursors.  Step growth polymerizations of isocyanates and acids are used to give high molecular weight polyamides.[4]
These processes are solution reactions at elevated temperatures involving the evaluation of $CO_2$.  These conditions are not amenable for
RIM modification.

$$R - COOH + Ar - NCO \longrightarrow R - CO - NH - Ar + CO_2$$

STEP GROWTH

Anionic initiation can effect polymerization of mono-isocyanates at
low temperatures giving a chain growth polymer of nylon 1.[5]

$$B^- + Ar-NCO \longrightarrow \left[ N(Ar)-CO \right]_n$$

CHAIN GROWTH

These polymers show very little thermal stability, unzipping at even
moderate temperatures giving the starting monoisocyanate.  Amides can
be prepared from isocyanates by addition of enamines 1 across the C-N
double band.

    Enamines, 1, are easily prepared from available starting mate-
rials in high yields.[6]  Many of these compounds, are low viscosity
liquids miscible with diols or polyols used in current urethane RIM
formulations.  The enamine/isocyanate addition reaction is rapid and
requires no catalysis.

## Experimental

    Enamines are formed by the condensation of an aldehyde or ketone
with a secondary amine.[6]

$$R\text{-}CO\text{-}CHR_2R_3 \; + \; R_4R_5NH \; \rightleftharpoons \; \underset{R_1}{\overset{R_4R_5N}{>}}\!=\!\underset{R_2}{\overset{R_3}{<}} \; + \; H_2O$$

Generally the carbonyl compound and a 1.5 - 2.0 fold excess of the amine are refluxed in benzene or toluene using a water separator (Method A) or molecular sieves in a soxhlet extractor (Method B) to remove water as it is formed.  Yields in excess of 80% are often realized (Table I).

Products must be protected from moisture to prevent hydrolysis, regenerating the original starting materials.

Polymeric systems, whether via solution or RIM processes, require difunctional enamines and diisocyanates.  Several difunctional and higher functional enamine systems were investigated.

Beta-substituted, acyclic enamines, 1, readily incorporate one equivalent of isocyanate to give a β-amine-α, β-unsaturated amide 2. The rate of this reaction is generally a function of the basicity of the parent amines $R_4R_5NH$.  More basic (i.e., nucleophilic) parent amines give more reactive enamines.  The mono-adduct 2 can incorporate a second equivalent of isocyanate to give an amide/lactam product 3.

Addition of a second isocyanate moiety to 2 is a very slow reaction. Compound 2 is deactivated by the conjugated amide as well as sterically hindered.  A β,β-disubstituted, non-deactivated enamine 4, reacts to give the lactam product shown in under 1 hour at room temperature.

Even starting with very reactive enamines, the monoadducts 2 show incomplete reactions to amide/lactams 3 after 48 hours at ele-

Table I

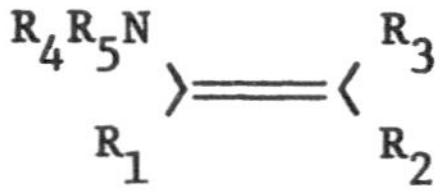

$$R_4R_5N \underset{R_1}{\overset{}{>}}\!\!=\!\!<\overset{R_3}{\underset{R_2}{}}$$

| $R_1$ | $R_2$ | $R_3$ | $R_4R_5N-$ | METHOD | YIELD |
|---|---|---|---|---|---|
| H | Et | Et | NMeAn | B | 67% |
| H | Et | Et | BNBA | B | 90% |
| H | Et | Et | PYR | A | 27% |
| H | | | | B | 76% |
| H | Et | H | PYR | B | 66% |
| $(CH_2)_3$ | | H | MOR | A | 82% |
| $(CH_2)_3$ | | H | NMeAn | A | 23% |
| $(CH_2)_4$ | | H | MOR | A | 85% |
| $(CH_2)_4$ | | H | NMeAn | A | 36% |

NMeAn–    N–Methylanilino–
DNBA–     Di–N–Butylamino–
PYR–      Pyrrolidino–
MOR–      Morpholino–

vated temperatures.

Since the addition of a second isocyanate to acyclic enamines is slow, $\alpha,\omega$-bisenamines will give bisamide products with isocyanates without the complication of further reaction to give crosslinking in polymeric systems.

Use of either $\alpha,\omega$-diamines or dicarbonyl compounds will give the appropriate $\alpha,\omega$-bisenamines.

$$R_4N\text{——}NR_4 \cdot R_1COCH_2R_2 \longrightarrow R_2CH{=}CR_1\text{-}NR_4\text{——}R_4N\text{-}CR_1{=}CHR_2$$

$$R_1COCH_2\text{——}CH_2COR_1 \cdot R_4R_5NH \longrightarrow {R_5R_4N \atop R_1}{\Large\rangle}{=}CH\text{——}CH{=}{\Large\langle}{NR_4R_5 \atop R_1}$$

Enamines derived from cyclic ketones have been shown to react difunctionally with isocyanates.[7]

Reactivity of these enamines is a function of ring size and the basicity of the parent amine. As previously noted, more basic parent amines give more reactive enamines. Cyclopentenyl systems are more reactive that cyclohexenyl systems. Cycloheptenyl systems are not reactive.

Incorporation of a second equivalent of isocyanate requires the rearrangement of the conjugated monoadduct 6 to the unconjugated adduct 7. Systems derived for less basic (i.e., aromatic) amines give high yields of only monoadduct 6 with one equivalent of isocyanate and reduced yields of the diadduct 8 in the presence of excess isocyanate. A rearrangement via intermediate 9 can accommodate the results.

More basic amine moieties facilitate rearrangement via zwitterion 9.

Thermal Stability of Products

The systems investigated show low to moderate thermal stabilities. Beta lactams, i.e., 10, show very low stabilities. Heating
compound 10 under vacuum at 100°C gave a quantitative recovery of
the starting enamine 11.

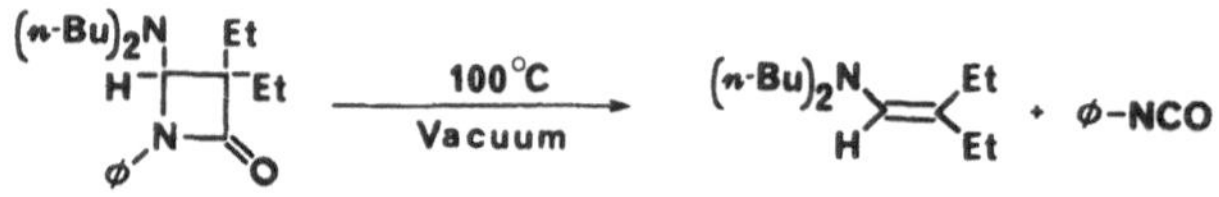

The thermal stabilities of amide products can be correlated with the
reactivities of the starting enamines. More reactive enamines give
products with lower thermal stabilities. This correlation implies
that the thermal degradation is simply the reverse of the amide-
forming reaction.

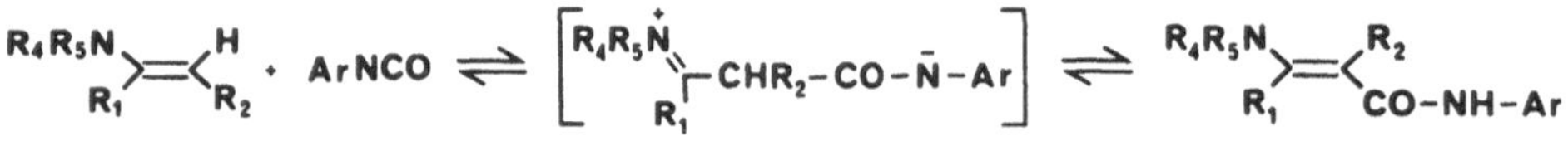

Factors that facilitate amide-forming reactions, ring size and amine
functionality, also facilitate the retro reaction. TGA analysis of
model compounds illustrates this trend. Model compounds 12-15
were heated from room temperature at 10°C/minute under nitrogen.
The weight loss at 225°C for each is shown.

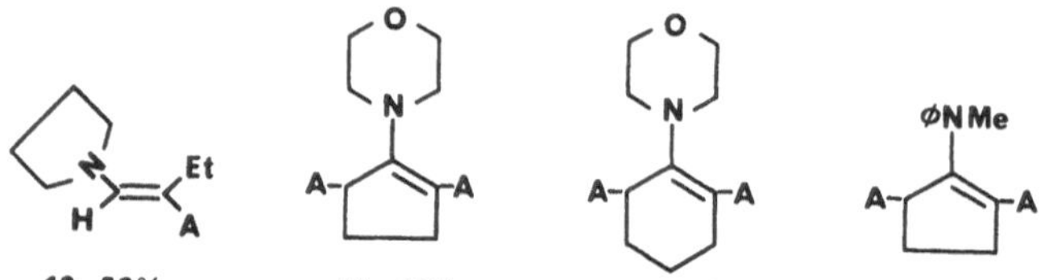

A = CONHφ

Conclusions

An examination of the chemistry of enamine/isocyanate reactions
for potential RIM use gave the following conclusions. Acyclic, mono-
enamines react too slowly (difunctionally) for RIM use. Additionally the amide/β-lactam products have unacceptably low thermal stabilities. Bisenamines react readily but are more difficult and ex-

pensive to prepare and purify.  Also, these compounds give products
with lower thermal stabilities than products derived from cyclic
enamines.  The cyclic enamines are easily prepared in good yields
from available starting materials.  These compounds show good reactivity with isocyanates at room temperature.  Reactivity can be altered by changes in ring size and/or amine functionality.


FORMULATIONS AND PROPERTIES

     Solution polymers employing difunctional enamines and isocya-
nates have been reported.[7]  An example is the reaction of 4,4'-meth-
ylene bis(isocyanato-benzene), MDI, with N-morpholino-1-cyclopentene
16.  The relevant chemistry involved has been previously discussed.
For ease of comparison, enamine 16 will be used in all subsequent
formulations.

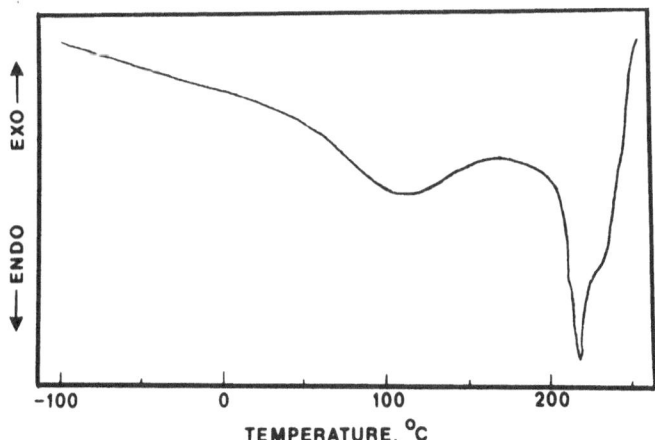

This polymer was prepared in dimethylacetamide at 50°C to an inherent
viscosity (m-cresol) of 0.60.  A DSC analysis of 4 (Figure 1) on a
duPont 910/990 system shows a glass transition at 75°C and a melt at
215°C.  Annealing a DSC sample at 150°C followed by a rescan elim-
inates the 75°C transition with an enhanced melt.  Thermogravimetric
analysis at 10°C/minute under nitrogen using a duPont 950/900 system
shows marked polymer degradation after 225°C.  Figure 2 gives the TGA

Figure 1.   DSC analysis of Polyamide 17

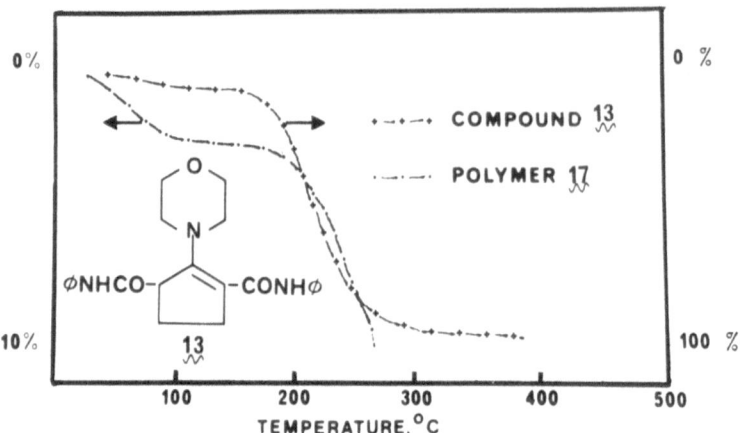

Figure 2.   Thermogravimetric analysis of Polyamide 17 and a Model
            Compound 13.


curves for the non-polymeric model compound 13 and polymer 17.   Thermal stabilities of these systems were previously discussed.[1]

## Experimental

Compound 16 is soluble in many of the polyols currently used in
urethane RIM technology.   Enamine 16 was therefore used as a urethane
chain extender ("hard segment") in concert with higher equivalent-
weight polyols ("soft segments").   Again for ease of comparison Thanol
SF-6503*, a six thousand molecular weight triol supplied by The Jef-
ferson Chemical Co., and Isonate 143L*, a liquid MDI-type isocyanate
supplied by The Upjohn Company, were used exclusively in all formulations discussed here.

Both isocyanate-capped prepolymers and "one shot" systems were
investigated.   Prepolymers were prepared from Isonate 143L* and
Thanol SF-6503*.   Parts were prepared by mixing an appropriate amount
of enamine with the prepolymer to give a desired isocyanate to extender equivalents ratio (i.e., the isocyanate index).   Systems at
lower isocyanate indices (1.00-1.05) required no catalysts.   Higher
index systems (1.15-1.50) incorporating isocyanurate moieties into
the final part required the use of an isocyanate trimerization catalyst.   Catalyst T-45*, supplied by M & T Chemicals, was used in all
these formulations.

"One shot" systems required preliminary preparation of a "B
Side".   A desired amount of enamine extender, Thanol SF-6503* and

---

*Registered trademarks.

a urethane catalyst T-12*, supplied by M & T Chemicals, were mixed together. Additionally, higher index systems required the addition of T-45* to the B side mixture. An appropriate amount of B side was then added to Isonate 143L* ("A side") to make parts.

Parts were prepared using a "hand mix" technique. Two hundred fifty grams of material, A plus B sides, were mixed together, rapidly stirred for four to five seconds and then cast into a 8" x 8" x 1/8" heated mold. The mold was then rapidly closed. In general these systems had gel times of 8-10 seconds after the onset of rapid mixing.

A mold temperature of 150°F (65.5°C) and a mold time of 2 minutes were routinely used. Parts were post cured at 250°F (121°C) for 1 hour after demold.

Mechanical properties of parts were determined using standard ASTM procedures. Heat sag measurements reflect the vertical drop in inches of a four inch cantilevered specimen after treatment at a given temperature for a given time.

Percentage of "hard segment" in the final piece reflects the total weight of extenders plus the weight of isocyanate that reacts with the extenders divided by the weight of the total system. Hard segment levels of 50% to 70% were investigated.

## Properties of Systems

Table II summarizes properties of hand mix systems prepared at a 50% hard segment level. Systems A and B use enamine 16 as the sole extender. System A employs a prepolymer; B is a one shot process. A comparable one shot, machine prepared system extended with ethylene glycol, system E, has a much greater tensile strength and better elongation. System B has a higher room temperature flex modulus and better thermal properties (heat sag). Attempts to realize the flexural and thermal properties of B and the tensile properties of E by using a mixture of extenders, systems C and D failed. Reasons for the poor tensile properties of these enamines-extended systems will be discussed. A DSC analysis (Figure 3) of system B shows a sharp soft segment glass transition at -55°C indicating good phase separation. A melt at 190-220°C and decomposition above 225°C are also seen. A TGA curve (Figure 4) indicates that thermal decomposition of system B parallels that of polymer 4 and model compound 5 (Figure 2).

Table III summarizes properties of systems at 65-67% hard segment levels. System F contains only amide hard segments; K is a

---

*Registered trademark.

Table II

1.05 NCO Index - 50% Hard Segment Systems

|                              | A     | B     | C     | D     | E     |
|------------------------------|-------|-------|-------|-------|-------|
| Shore D Hardness             | 55    | 55    | 55    | 55    |       |
| Tensile at Break, psi        | 1400  | 1600  | 1675  | 2450  | 3440  |
| % Elong. at Break            | 44    | 75    | 96    | 82    | 270   |
| Room Temp. Flex. Mod., psi   | 32000 | 26000 | 28000 | 29000 | 21000 |
| Heat Sag (250°F/1 hr.), in.  | 0.25  | 0.03  | 0.29  | 0.27  | 0.25  |
| Hard Segment - Amide         | 50    | 50    | 10    | 40    | 0     |
| - Urethane                   | 0     | 0     | 40    | 10    | 50    |

A - Thanol SF-6503/Isonate 143L prepolymer used.
B - One shot.
C & D - One shot using ethylene glycol coextender.
E - One shot using ethylene glycol-machine made.

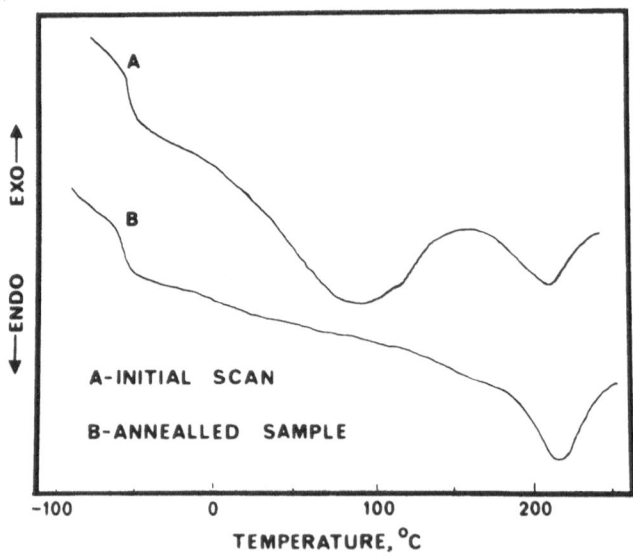

Figure 3.   DSC analysis of polyurethane - amide RIM system.

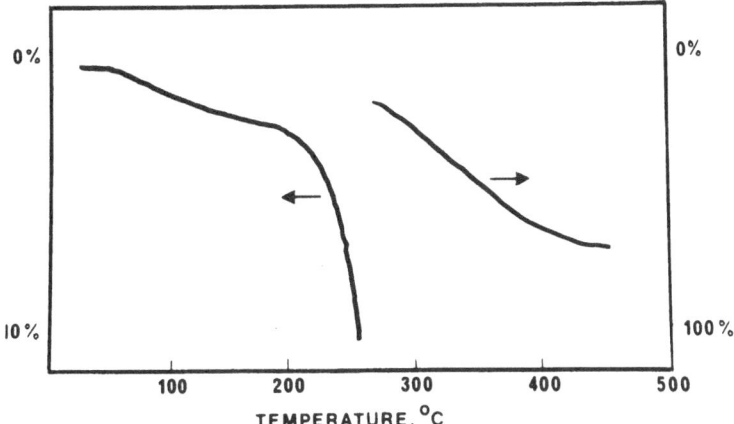

Figure 4.   Thermogravimetric analysis of polyurethane amide RIM
            system.

comparable machine-made ethylene glycol system.   These systems have
comparable thermal properties.   The ethylene glycol system has a
much greater elongation at break with about 60% of the room temperature flexural modulus of system F.   Use of 3,5-diethyl-2,4-diamino
toluene (diethyl TDA) as a coextender G, did not improve tensile
properties.   Systems H, I and J use low levels of trifunctional isocyanurate structures as coextenders.   Use of low levels of isocyanurate in ethylene glycol systems disrupts the hard segment crystallinity sufficiently to give pieces that distort badly at even post
cure temperatures.   Isocyanate index systems of 1.15, 1.25 and 1.50
have increasingly enhanced tensile, flexural, thermal and impact
properties.   Enamine-extended systems at this hard segment level
can all be characterized as having high flexural modulus, low impact
and good thermal (heat sag) properties.

    Table IV summarizes properties of systems that represent the best
balance of properties achieved using enamine 3.   A lower hard segment level, 55% coextended with low levels of isocyanurate resulted
in the best combination of tensile, flexural, thermal and impact
properties.

Conclusions

    Polyamide – RIM systems can be prepared from the reactions of
enamines with isocyanates.   These reactions may involve prepolymer
or one shot processes.   Enamines used are miscible in higher molecular weight polyols currently used in urethane RIM technology allowing formulation to a given hard segment level with a compatible B
side.   No catalysts are needed for the enamine/isocyanate reaction
used.

Table III

65% Hard Segment Systems

|  | F | G | H | I | J | K |
|---|---|---|---|---|---|---|
| Isocyanate Index | 1.05 | 1.05 | 1.15 | 1.25 | 1.50 | 1.05 |
| Shore D Hardness | 72 | 73 | 77 | 77 | 77 | 70 |
| Tensile at Break, psi | 3100 | 2950 | 4450 | 6400 | 6800 | 3600 |
| % Elong. at Break | 18 | 9 | 16 | 19 | 27 | 135 |
| Room Temp. Flex Mod., psi | 182000 | 155000 | 187000 | 190000 | 195000 | 109000 |
| Heat Sag in Inches: |  |  |  |  |  |  |
| 250°F/1 hr. | 0.02 | 0.00 | 0.02 | 0.02 | 0.02 | 0.02 |
| 325°F/0.5 hr. | 0.23 | 0.10 | 0.40 | 0.14 | 0.12 | 0.20 |
| Gardner Impact, ft. lb. | 0.50 | 0.20 | 0.70 | 0.60 | 0.80 |  |
| % Hard Segment |  |  |  |  |  |  |
| Amide | (67) 67 | (65) 47 | (67) 60 | (67) 60 | (65) 50 | 64 |
| Urea | -- | 18 | -- | -- | -- | -- |
| Trimer | -- | -- | 7 | 7 | 15 | -- |

F – Amide hard segment.
G – DETDA coextended.
H, I, J – Isocyanurate coextended.
K – Machine-made ethylene glycol system.

Systems extended with enamine 16 generally can be characterized as having high room temperature flexural modulus, good thermal properties, but poor tensile and impact properties. These properties reflect the crystalline hard segment. Coextension with low levels of isocyanurate provides the most effective remedy to the low impact properties.

Low tensile strengths most likely reflect the poor impact strengths of many of these systems. Test specimens simply break before reaching an ultimate tensile value. Low impact strengths may additionally reflect very early phase separation of these systems upon formation resulting in failure of the extender to react completely. Tensile and impact properties are often those most sensitive to polymer molecular weight.

Table IV

55% Hard Segment Systems

|  | L | M | N |
|---|---|---|---|
| Isocyanate Index | 1.15 | 1.25 | 1.50 |
| Shore D Hardness | 67 | 67 | 67 |
| Tensile at Break, psi | 3400 | 5100 | 5280 |
| % Elong. at Break | 90 | 80 | 60 |
| Room Temp. Flex Mod., psi | 126000 | 112000 | 123000 |
| Heat Sag in Inches: | | | |
| 250°F/1 hr. | 0.00 | 0.00 | 0.02 |
| 325°F/0.5 hr. | 0.55 | 0.97 | 0.60 |
| Gardner Impact, ft. lb. | 3.70 | 2.20 | 1.90 |
| % Hard Segment | 56 | 56 | 55 |
| Amide | 52 | 51 | 44 |
| Trimer | 4 | 5 | 11 |

References

1. P. S. Carleton, D. J. Breidenbach and L. M. Alberino, SPE NATEC, November, 1979.
2. G. Ferber, SPE NATEC, November, 1979.

3.  R. S. Kubiak and R. C. Harper, SPE NATEC, November, 1979.

4.  J. T. Chapin, B. K. Onder and W. J. Farrissey, Polymer Preprints, $\underline{21}$ (2), 130 (1980).

5.  Enamines: Synthesis, Structure and Reactions, A. Gilbert Cook, Dekker (1969).

6.  G. Stork et al, J. Am. Chem. Soc., $\underline{85}$, 207 (1963).

7.  W. H. Daley and W. Kern, Makromol. Chem., $\underline{108}$, 1 (1967).

# THE CATALYSIS OF THE POLYCYCLOTRIMERIZATION OF

# ISOCYANATES BY QUATERNARY AMMONIUM CARBOXYLATES

I. S. Lin*, J. E. Kresta and K. C. Frisch

Polymer Institute
University of Detroit
Detroit, MI 48221

## INTRODUCTION

Isocyanates can form under certain conditions cyclic trimers or linear polymers. Shashoua et al[1] have found that catalysis and reaction temperature are key factors which determine the composition of the final products. They observed that the formation of linear polymers proceeded only at low temperatures (< -20°C) and the formation of the cyclic trimers proceeded at ambient or higher temperatures.

The polycyclotrimerization of difunctional isocyanates (or NCO-terminated prepolymers) produces heterocyclic perhydro-1,3,5-triazine-2,4,6-trione (isocyanurate) rings as crosslinks:

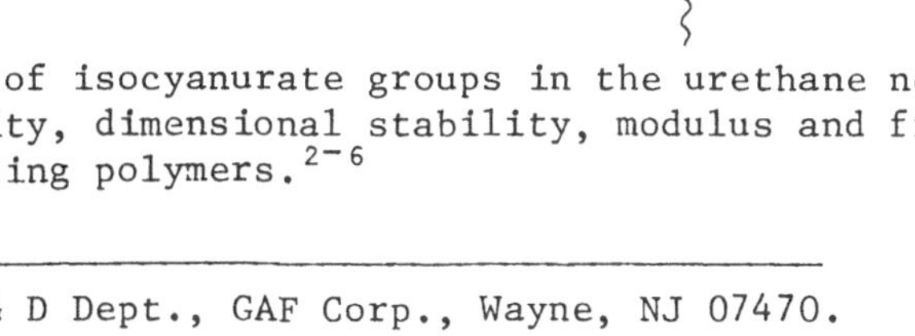

The presence of isocyanurate groups in the urethane network enhances thermostability, dimensional stability, modulus and fire resistance of the resulting polymers.[2-6]

---

*Plastics R & D Dept., GAF Corp., Wayne, NJ 07470.

The formation of the isocyanurate crosslinks in the urethane elastomer network can proceed simultaneously with the formation of the urethane linkages.  This is achieved when a higher isocyanate index is used during the reaction with an appropriate catalyst.  The high modulus isocyanurate-urethane elastomers can be potentially produced by the RIM technique.

Catalysis is a key to the successful preparation of isocyanurate-based copolymers.  Many potential catalysts have been reviewed in the literature.[4-7]  The substituted ammonium carboxylates[8] of the general formula

$$
\begin{array}{c}
R \\
| \\
R - N^+ - R' \qquad\qquad R''COO^- \\
| \\
R
\end{array}
$$

are very active cyclotrimerization catalysts with high solubility in prepolymers.  In this paper the results of the kinetic investigation of the structural effects on the cyclotrimerization activity of substituted ammonium carboxylates will be discussed.


EXPERIMENTAL

All chemicals used in this study were reagent grade.  Phenyl isocyanate (Aldrich Chemical Co., 99.6%) was purified by means of vacuum distillation.

## Purification of Solvents

Acetonitrile.  Acetonitrile was transferred in a dry three-necked flask and then calcium hydride (2 gm/liter) was added.  After the evolution of hydrogen, the acetonitrile was refluxed for four hours under a nitrogen blanket.  Subsequently, phosphorus pentoxide was added (3 gm/liter) and refluxing was continued for six hours. The acetonitrile was then distilled and the fraction with the constant boiling point (b.p. = 81.6°C) was collected and kept under nitrogen.

Ethyl Acetate.  Ethyl acetate was dried with anhydrous magnesium sulfate, then poured into a dry three-necked flask and was distilled under dry nitrogen.  The fraction with the constant boiling point (b.p. = 126.1°C) was collected and kept under nitrogen.

Measurement of Kinetics of Cyclotrimerization.  The kinetic measurements were carried out in an apparatus consisting of a thermostated 250 ml three-necked flask equipped with a nitrogen inlet, stirrer, reflux condenser and a recording thermocouple.  The phenyl isocyanate solution (50 ml) was pipetted into the flask and was stirred magnetically while 50 ml solution of catalyst in acetonitrile was placed in a dropping funnel, which was attached to the flask and was heated by an electric jacket. When constant temperature was obtained both in the flask and in the dropping funnel, the catalyst solution was transferred to the flask and the whole mixture was mixed intensively. Two to three gram samples were taken at regular time intervals and the isocyanate content was determined by the n-butyl-amine titration method.[9]

RESULTS AND DISCUSSION

The cyclotrimerization catalytic activity of substituted ammonium carboxylates with regard to their structure was studied in the model system phenyl isocyanate-acetonitrile. The cyclotrimerization reaction order, the effects of the structure of the catalyst and the relative permitivity of the solvent system on the catalytic activity of substituted ammonium carboxylates will be discussed in the following sections.

The Order of the Cyclotrimerization Reaction

The kinetics of the cyclotrimerization of phenyl isocyanate (0.5 mole $\ell^{-1}$) using quaternary ammonium carboxylates as catalysts was measured at various temperatures. It was found that the kinetics of the reaction followed second order with respect to the isocyanate as measured by the disappearance of the isocyanate group (correlation coefficient $\geq 0.99$). The second order was observed over a wide range of the catalyst concentrations and a high degree of conversion (Figures 1-4). However, the order with respect to the catalyst concentration was found to be a complex one. The kinetic data showed that the reaction can be first or second order (Figures 5a-5d) with respect to the catalyst concentration, depending on the structure of the catalyst. (See Table I for the formulae of the catalysts.)

As can be seen from Figures 6 and 7, catalysts I and II containing reactive hydroxyl groups were better fitted by first order at high catalyst concentration while the second reaction order with respect to the catalyst concentration correlated better over a wide catalyst concentration range. In the case of quaternary ammonium carboxylates III and IX, the cyclotrimerization of the isocyanate was clearly first order with respect to the catalyst concentration (Figures 8 and 9). It is interesting to note that at low catalyst concentrations, the cyclotrimerization reaction was extremely slow but the rate progressively increased with an increase of the catalyst level. Surprisingly, these low catalyst concentrations all fall into the initial bending curve region if we plot rate constant vs. catalyst concentration. This phenomenon can be due to the interaction of catalyst with the small amount of impurities present in isocyanate

or can be associated with the consecutive reaction of the hydroxyl
group of the catalyst with isocyanate.  The exact reason for this
catalyst behavior is not know at the present time.  The kinetic data
for catalysts III and IX (second reaction order to monomer, first
order to catalyst) can be interpreted by the cyclotrimerization
mechanism, proposed previously by Kresta et al.[10]

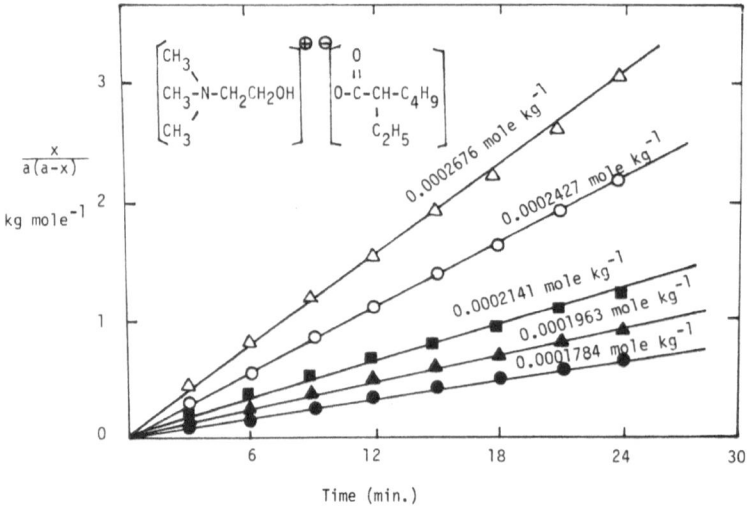

Figure 1.  Cyclotrimerization of phenyl isocyanate catalyzed by sub-
           stituted ammonium carboxylate I in acetonitrile at 30°C.

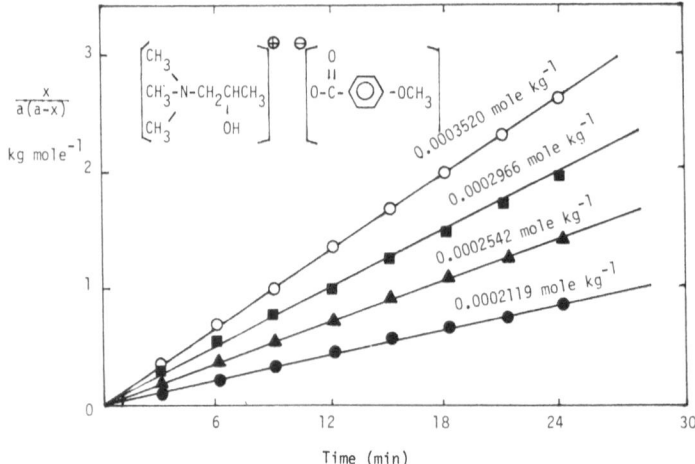

Figure 2.  Cyclotrimerization of phenyl isocyanate catalyzed by sub-
           stituted ammonium carboxylate IV in acetonitrile at 30°C.

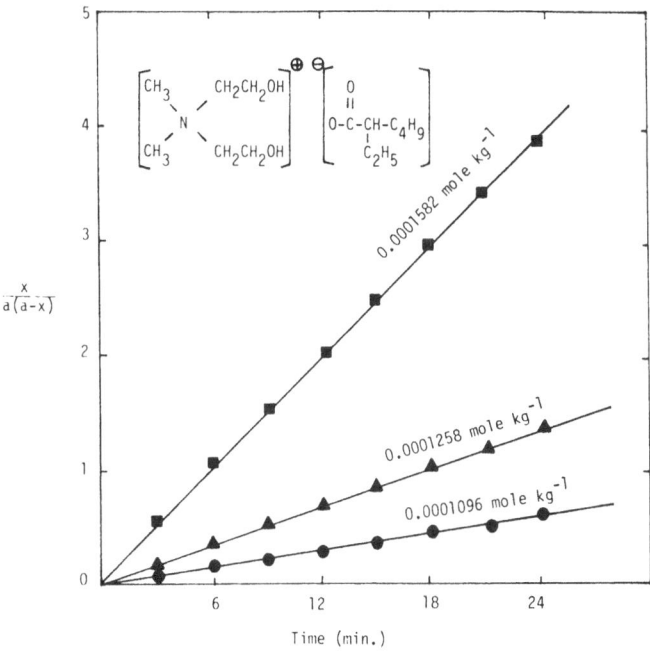

Figure 3.  Cyclotrimerization of phenyl isocyanate catalyzed by substituted ammonium carboxylate VII in acetonitrile at 30°C.


## Effects of the Structure of Quaternary Ammonium Carboxylates on Catalytic Activity

Nucleophilicity Effects of the Active Center.  It was found that
the catalytic activity of the ammonium carboxylates increased with
the increase of the nucleophilicity of the active center $-COO^{\ominus}$.  The
following sequence of decreasing catalytic activity

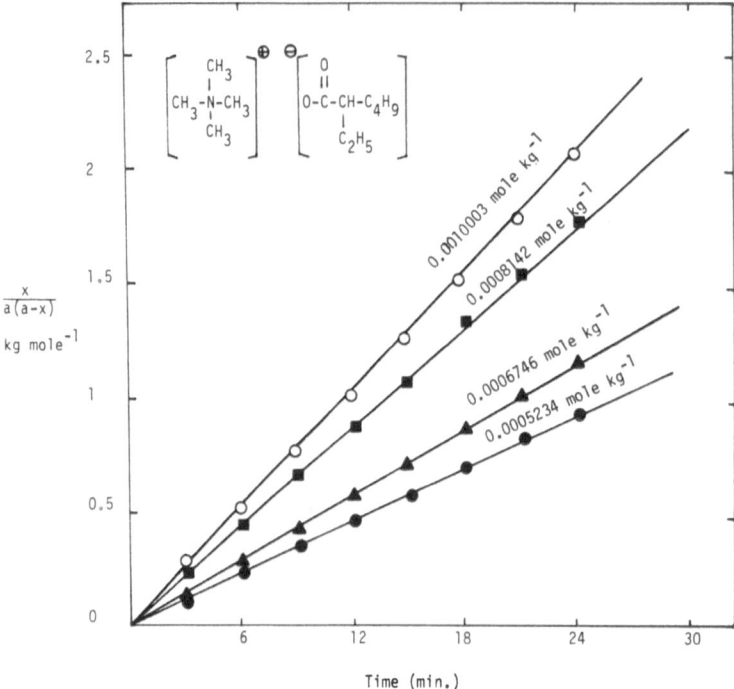

Figure 4.  Cyclotrimerization of phenyl isocyanate catalyzed by substituted ammonium carboxylate IX in acetonitrile at 30°C.

shows that electron donating substituents (-OCH$_3$) on the benzene
ring increased the k$_{cat}$ values and electron-acceptor (-Cl) decreased
the k$_{cat}$ values in the cyclotrimerization of phenyl isocyanate.  The
rate constants k$_{cat}$ were correlated with the sigma constants for
substituents (which are directly related to the changes in electron
densities on the reaction center -COO$^\ominus$), using the Hammett equation:

$$\log \frac{(k_{cat})_X}{(k_{cat})_H} = \sigma \rho$$

The results are presented in Figure 10.  As can be seen from Figure
10, there is direct correlation between the nycleophilicity of the
active center -COO$^-$ (determined by the donor-acceptor strength of
substituent X on the benzene ring) and the cyclotrimerization activity of the catalyst.  This correlation can be extended to the aliphatic substituted ammonium carboxylates by substitution of σ by log
K/K$_H$, where K and K$_H$ are ionization constants for acids.  The kinetic
data for II, IV, V and VI (with similar steric structures around the
nitrogen cation) correlated well according to the equation:

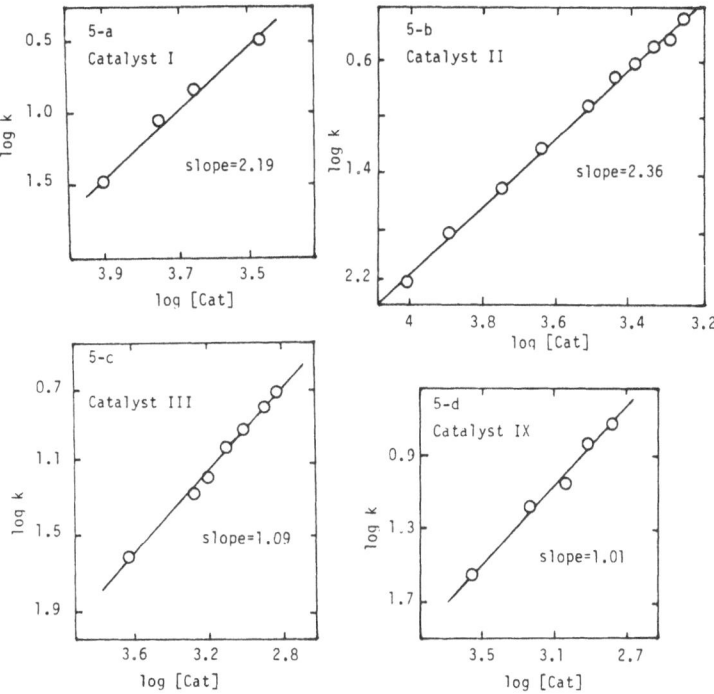

Figure 5.   The order of the cyclotrimerization reaction with
respect to the catalyst concentration.   T = 25°C.

$$\log k_{cat} = - \rho \ pKa$$

as can be seen from Figure 11.

**Effect of the Hydroxyl Group of the Cation on the Activity of
a Catalyst.**  The presence of the hydroxyl group on the ammonium
cation significantly increased the catalytic activity of the catalyst.[11],[12]  (See Table I).  This effect is probably associated with
the formation of a carbamate group bonded directly to the cationic
part of the catalyst:

$$- \overset{\mid}{\underset{\mid}{N}}^{\oplus} \!\!\! - CH_2CH_2\text{-}OH + RNCO \longrightarrow - \overset{\mid}{\underset{\mid}{N}}^{\oplus} \!\!\! - CH_2CH_2O\text{-}\overset{O}{\underset{\parallel}{C}}\text{-}NH\text{-}R$$

Kresta and Hsieh found that the urethane group is a strong co-
catalyst in the cyclotrimerization reaction of isocyanate.[13],[14]
This co-catalytic effect of the urethane group is probably associated
with induced polarization of the isocyanate groups through the formation of hydrogen bonding.

Table I.  Cyclotrimerization of Isocyanates.  Relative Catalytic
Activity of Substituted Ammonium Carboxylates.  T = 30°C.
*) pseudo first order to catalyst.

| Catalyst | Relative Catalytic Activity $k_{cat}/k_{cat}$ IX |
|---|---|
| I. $[(CH_3)_3NCH_2CH_2OH]^{\oplus}$  $^{\ominus}[OOC\ CH(C_2H_5)C_4H_9]$ | 10.76 *) |
| II. $[(CH_3)_3NCH_2CH(CH_3)OH]^{\oplus}$  $^{\ominus}[OOCCH(C_2H_5)C_4H_9]$ | 6.87 *) |
| III. $[(CH_3)_3NCH_2C(CH_3)_2OH]^{\oplus}$  $^{\ominus}[OOCCH(C_2H_5)C_4H_9]$ | 1.17 |
| IV. $[(CH_3)_3NCH_2CH(CH_3)OH]^{\oplus}$  $^{\ominus}[OOC\text{-}\bigcirc\text{-}OCH_3]$ | 5.11 *) |
| V. $[(CH_3)_3NCH_2CH(CH_3)OH]^{\oplus}$  $^{\ominus}[OOCC_6H_5]$ | 4.45 *) |
| VI. $[(CH_3)_3NCH_2CH(CH_3)OH]^{\oplus}$  $^{\ominus}[OOC\text{-}\bigcirc\text{-}Cl]$ | 3.74 *) |
| VII. $[(CH_3)_3N(CH_2CH_2OH)_2]^{\oplus}$  $^{\ominus}[OOCCH(C_2H_5)C_4H_9]$ | 28.01 *) |
| VIII. $[(CH_3)_3N(CH_2CH_2OH)_3]^{\oplus}$  $^{\ominus}[OOCCH(C_2H_5)C_4H_9)]$ | -- |
| IX. $[(CH_3)_4N]^{\oplus}$  $^{\ominus}[OOCCH(C_2H_5)C_4H_9]$ | 1.00 |
| X. $[(CH_3)_3NCH_2CH_2OH]^{\oplus}$  $^{\ominus}[OOCCH_3)$ | 10.00 *) |
| XI. $[(CH_3)_3NCH_2CH_2OH]^{\oplus}$  $^{\ominus}[OOCH]$ | 9.34 *) |

$$\sim\sim\sim \underline{N} = C = \underline{O}1 \overset{\Delta+}{} \overset{\Delta-}{} \cdots\cdots H\text{-}N$$

with $C = O$ group attached to N.

The increased electrophilicity of the carbon atom of the isocyanate
group made initiation (nucleophilic attack of the NCO group by the
catalyst) and propagation reaction easier and, as a result, a higher
reaction rate of cyclotrimerization was observed.

The relative activity of substituted ammoniun carboxlates containing primary, secondary and tertiary hydroxyl groups decreased in
the following order:

$$\left[(CH_3)_3NCH_2CH_2\underset{OH}{\overset{}{|}}\right]^{\oplus} \left[O\text{-}\overset{O}{\overset{\|}{C}}\text{-}CH\underset{C_2H_5}{\overset{}{|}}\text{-}C_4H_9\right]^{\ominus} > \left[(CH_3)_3NCH_2\overset{CH_3}{\overset{|}{CH}}\underset{OH}{}\right]^{\oplus} \left[O\text{-}\overset{O}{\overset{\|}{C}}\text{-}CH\underset{C_2H_5}{\overset{}{|}}\text{-}CH_4H_9\right]^{\ominus} \gg$$

$$\left[ (CH_3)_3NCH_2 \overset{\overset{\displaystyle CH_3}{|}}{\underset{\underset{\displaystyle OH}{|}}{C}}-CH_3 \right]^{\oplus} \left[ O-\overset{\overset{\displaystyle O}{\|}}{C}-\overset{}{\underset{\underset{\displaystyle C_2H_5}{|}}{CH}}-CH_4H_9 \right]^{\ominus}$$

Relative reactivity:          I    >    II    >>    III
                             1.0       0.6         0.1

The results are in agreement with the expected influence of steric hindrance due to branching of the aliphatic chain close to the site of reaction.   The branching interferes with the carbamate formation resulting in the weakening of the co-catalytic effect.   Consequently the rate of trimerization is lowered as the branching of the side chain increases.

A similar order of relative reactivity (see Table II) was determined for the urethane reaction where primary, secondary and tertiary alcohols were reacted with isocyanate in benzene at 26°.[15]

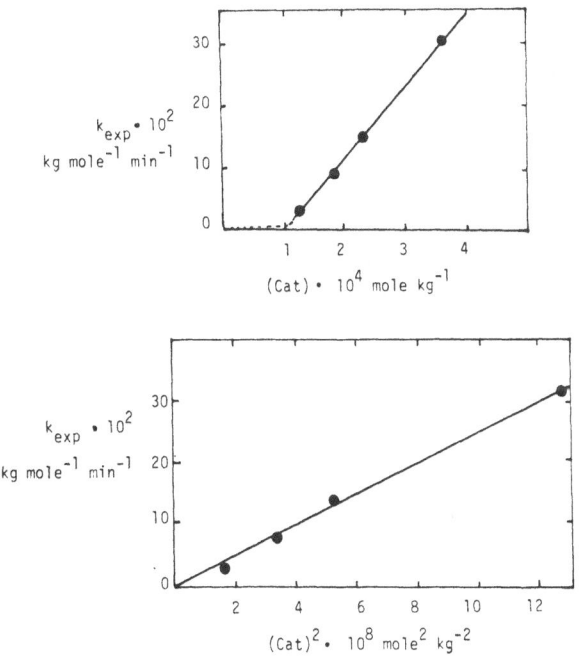

Figure 6.   Relationship between $k_{exp}$ and concentration of substituted ammoniun carboxylate I.   T = 25°C.

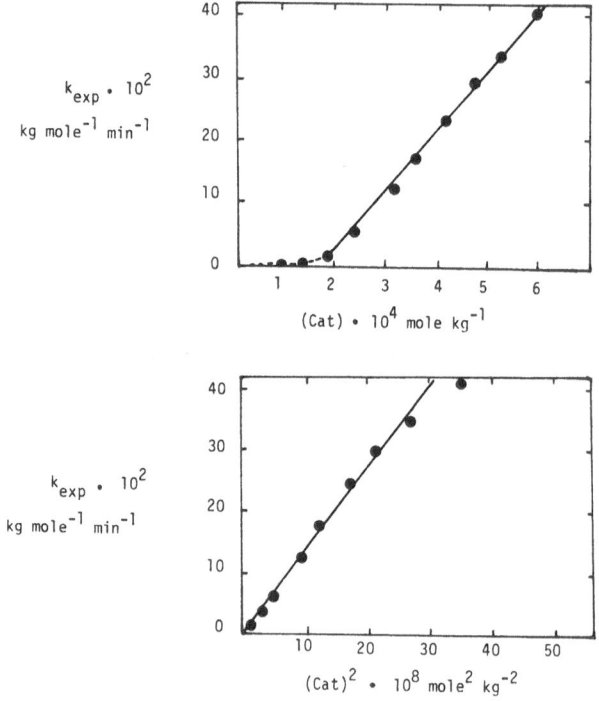

Figure 7.   Relationship between $k_{exp}$ and concentration of
            substituted ammonium carboxylate II.   T = 25°C.

|  | primary<br>alcohol | > | secondary<br>alcohol | > | tertiary<br>alcohol |
|---|---|---|---|---|---|
| Relative reactivity: | 1.0 | | 0.3 | | 0.003-0.009 |

It is interesting to note that a similar relationship was observed
in the case of aryl aminimide catalysts.[16]

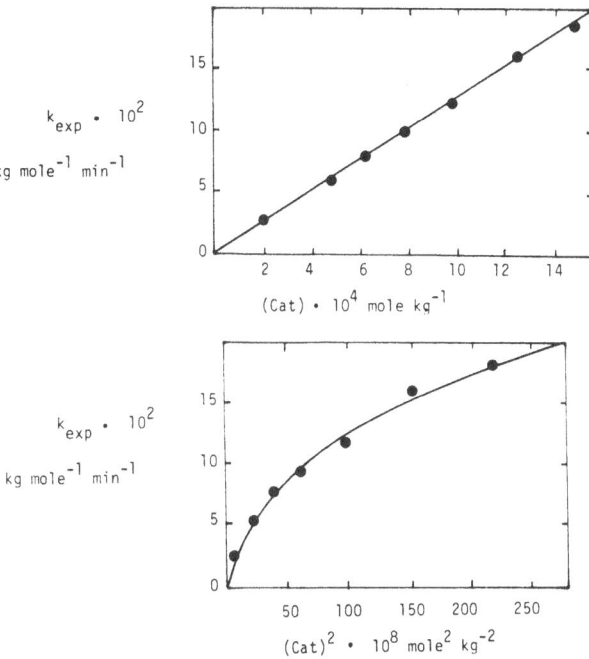

Figure 8.   Relationship between $k_{exp}$ and concentration of
            substituted ammonium carboxylate III.   T = 25°C.


__Effects of Functionality (Number of Hydroxyl Groups Bonded to__
__the Nitrogen Cation) on Catalytic Activity.__  It was found that the
catalyst with two hydroxyl groups beonded to the nitrogen cation
(catalyst VII) exhibited much higher catalytic activity than catalyst
I, containing only one hydroxyl group (see Table I).

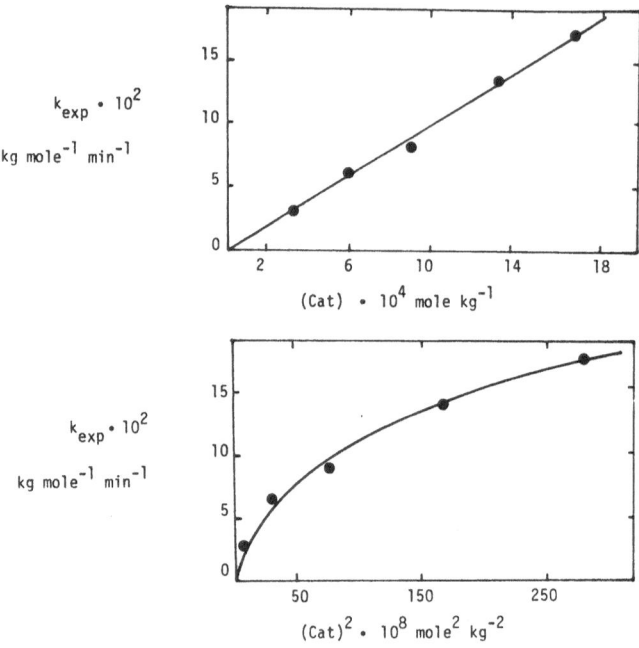

Figure 9.   Relationship between $k_{exp}$ and concentration of substi-
            substituted ammonium carboxylate IX.   T = 25°C.

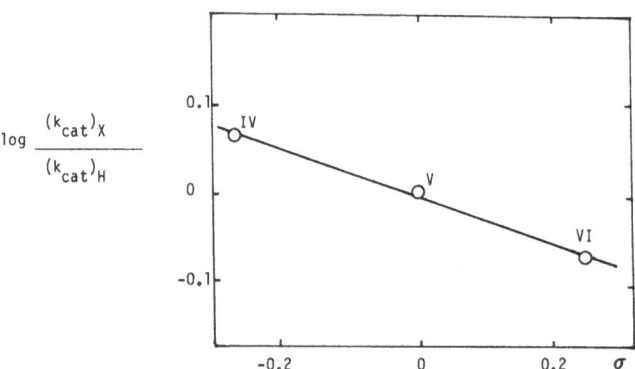

Figure 10.   Cyclotrimerization of isocyanates.   Correlation of
             log $[(k_{cat})_x]/[k_{cat})_H]$ with the Hammett sigma parameter
             of substituted phenyl isocyanates.   T = 30°C.

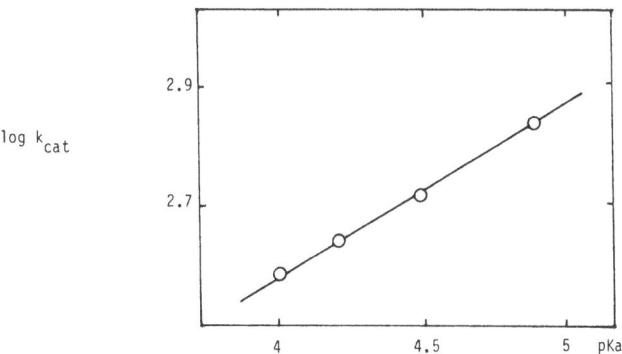

Figure 11.  Cyclotrimerization of isocyanates correlation between
log $k_{cat}$ and pKa values of acids of substituted ammonium
carboxylate catalysts II, IV, V, VI.   T = 30°C.

Table II.   Relative Reactivities of Aliphatic
Alcohols with Phenyl Isocyanate[*]

| Name of Alcohol | Chemical | Relative Rate of Reactivity |
|---|---|---|
| Ethanol | $CH_3CH_2OH$ | 302 |
| 2-Propanol | $CH_3CH\text{-}OH$<br>$\quad\ CH_3$ | 96 |
| 2-Methyl-1-propanol | $\qquad CH_3$<br>$CH_3C\text{-}OH$<br>$\qquad CH_3$ | 1 |
| 1-Propanol | $CH_3CH_2CH_2OH$ | 108 |
| 2-Butanol | $CH_3CH_2CH\text{-}OH$<br>$\qquad\ \ CH_3$ | 44 |
| 2-Methyl-2-butanol | $\qquad\ CH_3$<br>$CH_3CH_2C\text{-}OH$<br>$\qquad\quad CH_3$ | 1 |

[*]David, T.L. and Farnum, J. M., J. Am. Chem. Soc., **56**, 883 (1934).

However, catalyst VIII with three hydroxyl groups showed low activity
(Figure 12). One possible explanation is that the steric hindrance
and bulkiness of the branched cation is interfering in the propagation step of cyclotrimerization and the reaction does not achieve
completion.[19] This phenomenon can be practically utilized for the
preparation of partially trimerized diisocyanates. These partially
trimerized diisocyanates containing isocyanurate rings can be utilized in the preparation of elastomers, coatings and foams.[20]

### Solvent Effect

The effect of relative permitivity on the cyclotrimerization
reaction was studied on the solvent system based on mixtures of acetonitrile and ethyl acetate. The quaternary ammonium carboxylate
III was used as catalyst. It was observed that the rate constant
$k_{cat}$ increased with the increase of the relative permitivity of the
solvent system. The experimental data correlates well with the
Kirkwood equation (see Figure 13) which supports the idea of the
ionic nature of the cyclotrimerization reaction. However, the overall stiuation is complicated by the fact that the reaction is also
affected by the solvation of reactants by solvents, as was demonstrated by the kinetic measurements in solvent systems having the
same relative permitivity.[20,21]

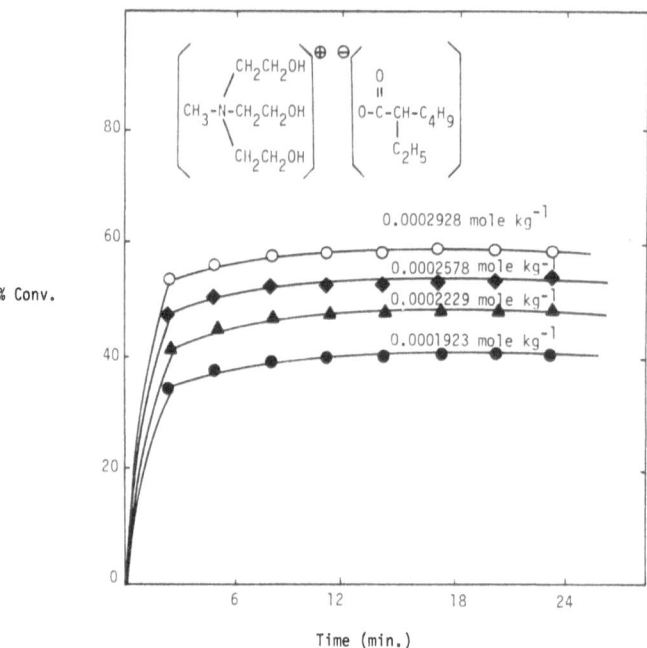

Figure 12. Cyclotrimerization of phenyl isocyanate catalyzed by quaternary ammonium carboxylate VIII in acetonitrile at 40°C.

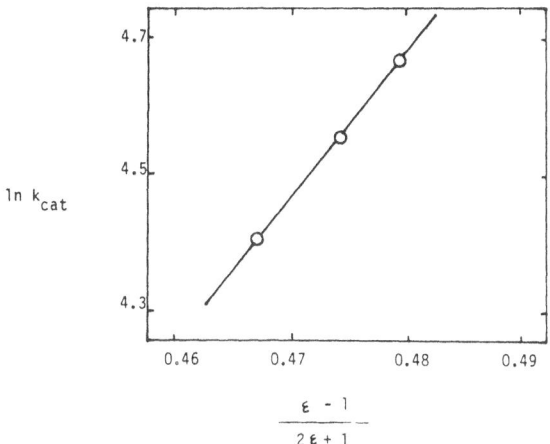

Figure 13.   Effect of relative permitivity of solvent system on cyclotrimerization reaction.  Solvent system:  acetonitrile/ethyl acetate.  Catalyst:  III.

$[(CH_3)_3NCH_2C(CH_3)_2OH]^{\oplus} \quad {}^{\ominus}[OOCCH(C_2H_5)C_4H_9]$

T - 25°C

SUMMARY

The kinetics of the cyclotrimerization of phenyl isocyanate by substituted ammonium carboxylate catalysts followed second reaction order with respect to the isocyanate.  The order with respect to the catalyst was of first or second power depending on the structure of the ammonium carboxylate.  The second order with respect to concentration of the catalyst was typical for ammonium carboxylates containing hydroxyl groups capable of reacting with isocyanate.

The cyclotrimerization activity of substituted ammonium carboxylates increased with the increase of the nucleophilicity of the active center (COO⁻).

The substituted ammonium carboxylates containing primary hydroxyl groups exhibited greater reactivity than substituted ammonium carboxylates containing secondary or tertiary hydroxyl groups.

The substituted ammonium carboxylates containing two hydroxyl groups on the nitrogen atom were more reactive than ammonium carboxylates containing one hydroxyl group.  This was valid only for systems based on monofunctional isocyanates.

The cyclotrimerization activity of substituted ammonium carboxylates increased with the increase of the relative permitivity of

the solvent system.  This phenomenon supports the idea of the ionic
nature of the cyclotrimerization reaction.

## Acknowledgements

The financial support of Air Products and Chemicals, Inc., and
of the National Science Foundation is gratefully acknowledged.

## References

1.  Shashoua, V. E., Sweeny, W., Tietz, R. F, J. Am. Chem. Soc.,
    $\underline{82}$, 866 (1960).
2.  Nicholas, L., and Gmitter, G. R., J. Cellular Plast. $\underline{1}$, 85
    (1965).
3.  Sayigh, A. A. R., Advances in Urethane Science and Technology,
    Vol. III, 141 (1974), Technomic Publishing Co., Westport, CT.
4.  Nawata, T., Kresta, J. E., and Frisch, K. C., J. Cellular Plast.
    $\underline{11}$, 267 (1975).
5.  Frisch, K. C., and Kresta, J. E., International Progress in
    Urethanes, Vol. 1, 191 (1977), Technomic Publishing Co., West-
    port, CT.
6.  Sasaki, N., Yokoyama, T., Tanaka, T., J. Polym. Sci., A-1, $\underline{11}$,
    248 (1970).
7.  Robins, J., O'Keefe, D. R., Advances in Urethane Science and
    Technology, $\underline{8}$, 185 (1982), Technomic Publishing Co., Westport,
    CT.
8.  Bechara, I. S., Carroll, F. P., Holland, D. G., and Masciolli,
    R. L., U. S. Pat. 4,040,992 (1977).
9.  David, D. J., Staley, H. B., "Analytical Chemistry of Polyure-
    thanes," Wiley-Interscience, NY, 1969.
10. Kresta, J. E., Shen, C. S., Frisch, K. C., ACS Org. Coatings
    and Plastics Chemistry, $\underline{36}$, 2, 674 (1976).
11. Bechara, I. S., and Carroll, F. P., SPI, Atlanta (1978).
12. Kresta, J. E., Lin, I. S., Hsieh, K. H, and Frisch, K. C., ACS
    Org. Coatings and Plastics Preprints, $\underline{40}$, 911 (1979).
13. Kresta, J. E., Hsieh, K. H., Makromol. Chem. $\underline{179}$, 2799 (1978).
14. Kresta, J. E., Hsieh, K. H., Makromol. Chem. $\underline{180}$, 793 (1979).
15. David, T. L., and Farnum, J. M., J. Am. Chem. Soc., $\underline{56}$, 883,
    (1934).
16. Kresta, J. E., Chang, R. J., Kathiriya, S., and Frisch, K. C.,
    Makromol. Chem. $\underline{180}$, 1081 (1979).
17. Kresta, J. E., Lin, I. S., Shen, C. S., Hsieh, K. H., and
    Frisch, K. C., ACS Org. Coatings and Plastics Preprints, $\underline{39}$,
    540 (1978).
18. Kresta, J. E., Shen, C. S., and Lin, I. S., SPE Technical
    Paper $\underline{25}$, 435 (1979).
19. Bechara, I. S., and Masciolli, R. L., J. Cell. Plast., $\underline{15}$,
    102 (1979).

20.  Shen, C. S., Dissertation, University of Detroit, Chapt. V, 139 (1976).

21.  Kresta, J. E., Shen, C. S., and Frisch, K. C., Proceedings IUPAC Int. Symp. on Macromolecules, Dublin, Vol. I, 135 (1977).

# HIGH MODULUS ISOCYANURATE-URETHANE RIM ELASTOMERS

J. E. Kresta, K. H. Hsieh and C. L. Wang

Polymer Institute
University of Detroit
Detroit, MI 48221

## INTRODUCTION

Urethane elastomers are segmented block copolymers composed of polyether or polyester soft segments which are connected with the urethane hard segments, formed from diisocyanates and low molecular weight diols or diamines (chain extenders). The segments are incompatible at lower temperatures and therefore, microphase separation proceeds with the formation of soft and hard segment domains. The mechanical properties depend strongly on the phase structure, the content of the hard segments in elastomers, on the molecular weight, distribution and functionality of the soft segments. The degree of phase separation, the size of domains and crystallization play an important morphological role affecting overall properties of urethane elastomers.[1-6]

The morphology of RIM elastomers is also strongly affected by the thermal history, mixing conditions of reactants and by the crosslinking reaction.[7,8] The hard domains, which contain urethane groups hydrogen bonded to a high degree act as fillers and physical crosslinks and their content is a determining factor of the elastomer modulus. Therefore, the high modulus urethane elastomers are usually prepared from the graft or polymer reinforced polyether polyols, 4,4'-diphenyl methane diisocyanate (MDI) and from the chain extenders, used at high concentration levels.[17,18]

The chemical crosslinking of the hard segments and its effect on the mechanical properties at high temperatures was not systematically studied. In this paper we discuss properties of elastomers containing crosslinked hard segments via isocyanurate groups.

165

It is well established that under certain reaction conditions isocyanates can form cyclic trimers or linear polymers. Shashoua et al[9] have found that catalysis and reaction temperature are key factors which determine the composition of the resulting reaction products. They observed that the formation of linear polymers proceeded only at low temperatures (< -20°C) and the formation of cyclic trimers at ambient or higher temperatures. Under certain steric conditions formation of polycyclic structures were observed by Butler et al[10,11] and Iwakura.[12]

The polycyclotrimerization of polyfunctional isocyanates (or NCO-terminated prepolymers) produces polymer networks containing heterocyclic perhydro-1,3,5-triazine-2,4,6-trione (isocyanurate) rings as crosslinks:

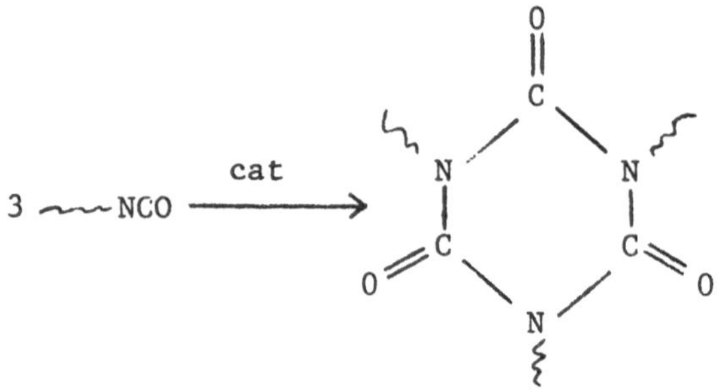

The isocyanurate rings are thermally stable (decomposition started at temperatures above 400°C[13]) and therefore cyclotrimerization of difunctional isocyanates can be utilized for preparation of thermally stable polymers.[14,15]

The formation of isocyanurate crosslinks in the elastomer network can proceed simultaneously to formation of the urethanes:

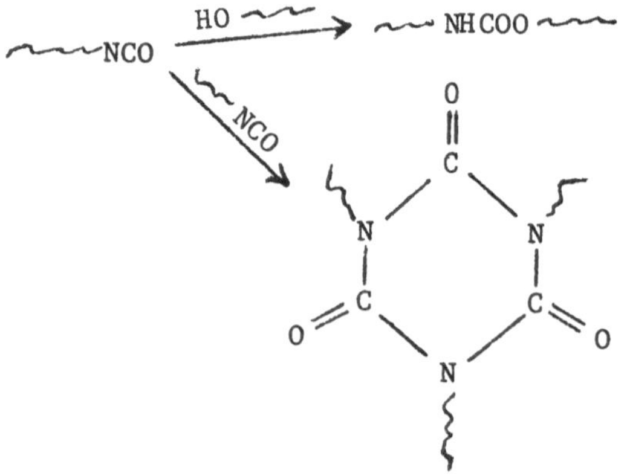

This is achieved when higher isocyanate index is used during reaction with an appropriate catalyst.[15] The isocyanurate-urethane elastomers can be potentially processed by the RIM technique. The properties of the resulting isocyanurate-urethane elastomers are a subject of this study.


EXPERIMENTAL

Chemicals

Methylene diphenyl diisocyanate (MDI, 99% +, from the Mobay Chemical Co.) and modified MDI (Isonate 191, isocyanate equivalent 139.5, from the Upjohn Chemical Co.) were used without further purification. Poly(oxytetramethylene diol) (POTMD, mol. wt. 1000 and 2000, from the Quaker Oats Chem. Co.) and grafted polyether polyol (Niax 31-28, Union Carbide, mol. wt. of the base polyol was 5000, content of acrylonitrile 21%, hydroxyl number 28) were used as polyols. 1,4-Butanediol (GAF) was used as a chain extender. Boron organometallic complex (Hexcel Corp.) and N,N',N"-tris(dimethylaminopropyl) sym. hexahydrotriazine (TDAPH) were used as catalysts.

Preparation of Isocyanurate-Urethane Elastomers

Model Isocyanurate-Urethane Elastomers. The model isocyanurate-urethane elastomers were prepared by the polycocyclotrimerization reaction of the NCO-terminated prepolymers with MDI at various molar ratios using TDAPH as catalyst. The casting technique (110°C, 17 hours) was used in the sample preparation. The NCO-terminated prepolymers were prepared by reacting MDI with poly(oxytetramethylenediol) (mol. wt. = 1000, 2000) in the reaction kettle, equipped with the nitrogen inlet and outlet, mechanical stirrer, heating mantle and a thermometer. The reaction was carried out at 70°C without any catalyst and was terminated when the NCO content reached the theoretical values.[16]

Lower Isocyanate Index Urethane-Isocyanurate Elastomers. The low isocyanate index urethane-isocyanurate elastomers were prepared from the graft polyether polyol (Niax 31-28, Union Carbide), modified 4,4'-diphenylmethane diisocyanate (Isonate 191, Upjohn Chem. Co., isocyanate equivalent 139.5), 1,4-butanediol (chain extender) and organometallic boron catalyst (Hexcel Corp.). The Isocyanate Index was varied from 150 to 250. The pre-dried polyol was blended with chain extender, catalyst and isocyanate. The blend was stirred for twenty seconds in a high speed mixer at room temperature. At the end of mixing the blend was degassed and immediately poured into the pre-heated mold and molded in the press for three minutes. Then the sample was removed and post-cured one hour at 120°C in an oven. The elastomers were then conditioned at room temperature (50% relative humidity) for one week prior to testing.

## Characterization of Isocyanurate-Urethane Elastomers

The infrared spectra of elastomers were recorded using a Unicam Infrared Spectrophotometer, Model 3-300. The dynamic viscoelastic properties of copolymers were measured by using Rheovibron Model DDV-II (Toyo Measuring Instruments Co.). The measurements were carried out at intervals of 1-2°C in the transition region (otherwise at 3-5°C) at a frequency of 110 Hz. The thermal analysis of the copolymers were carried out using a Perkin-Elmer Differential Scanning Calorimeter, Model DSC-2.

Stress-strain properties were determined at various temperatures by using an Instron Tensile Tester, Model 1113, with an environmental chamber, according to the ASTM D-412-75 test method. The flexural modulus was determined according to the ASTM D-790 method and the Shore D hardness according to the ASTM D-2240 method.

## RESULTS AND DISCUSSION

### Cyclotrimerization of Isocyanates

The previous investigations of the kinetics of cyclotrimerization of isocyanates revealed that the cyclotrimerization reaction proceeded through initiation, propagation, transfer and termination steps.[19] The reactivity of isocyanates in the cyclotrimerization reaction depends on the structure of isocyanate, nucleophilicity of a catalyst and the relative permitivity of the reaction system.[15] Isocyanates with electron withdrawing substituents adjacent to the NCO groups are more reactive than isocyanates with electron-donating groups. Therefore, aromatic isocyanates cyclotrimerize significantly faster than aliphatic or cycloaliphatic ones. (Table I). The kinetic data for the cyclotrimerization of substituted phenyl isocyanates correlated with the Hammett equation, giving the ρ value of +1.57. The large positive ρ value supports the idea that the cyclotrimerization reaction proceeds via attack of the nycleophile (a catalyst or a propagating chain) on the electrophilic carbon atom of the isocyanate group. It is interesting to note that the NCO-terminated prepolymers cyclotrimerize faster than the original isocyanates. The phenomenon is due to the co-catalytic effect of urethane groups which participate in the induced polarization of the isocyanate groups.[20,21]

The relative permitivity of the reaction system had a very strong effect on the cyclotrimerization rate; with the increase of the relative permitivity of the system, the cyclotrimerization rate constants increased. The experimental data[15] correlate well with the Kirkwood equation:

Table I.   Reactivity of Isocyanates in Cyclotrimerization Reaction

> Catalyst = Sodium ethoxide
>         T = 40°C
>   Solvent = DMF

| Isocyanate | $\dfrac{(k_{isoc})_{cat}}{(k_{BuNCO})_{cat}}$ |
|---|---|
| Prepolymer (MDI-POTMD 650) | 974.5 |
| p-ClPhNCO | 370.5 |
| PhNCO | 177.3 |
| MDI | 166.5 |
| p-MePhNCO | 86.5 |
| HDI | 3.85 |
| BuNCO | 1 |
| CHI | 0.41 |

$$\ln k_{cat} = \ln k_{cat_0} - \frac{1}{kT}\left(\frac{D-1}{2D+1}\right)\left[\frac{\mu^2_C}{r^3_C} + \frac{\mu^2_M}{r^3_M} - \frac{\mu^2_{CM}{}^{\ddagger}}{r^3_{CM}{}^{\ddagger}}\right]$$

where $\mu_C$, $r_C$, $\mu_M$, $r_M$, $\mu_{CM}{}^{\ddagger}$, $r_{CM}{}^{\ddagger}$ are dipole moments and radii for catalyst, isocyanate and activated complex CM‡, respectively.

The observed positive value for the second term indicated that the activated complex (created between monomer and catalyst, or propagating chain) had a larger separation of charges than the reactants in the initial stage and that the cyclotrimerization reaction was of the ionic nature.

## Preparation and Properties of Isocyanurate-Urethane Copolymers

Model isocyanurate-urethane copolymers (polycocyclotrimers) of variable crosslink density were prepared by the polycocyclotrimerization of difunctional isocyanates (MDI) with the NCO-terminated prepolymers of variable chain lengths and molar ratios, using TDAPH as catalyst. It was found that a comparable reactivity and mobility of comonomers and good solubility of a trimerization catalyst in

monomers (or monomer-solvent system) were desirable for the successful preparation of polycocyclotrimers. The formation of isocyanurate
rings in the polycocyclotrimerization reaction of isocyanates was
characterized by the disappearance of the NCO groups ($\nu$[N=C=O] 2280
$cm^{-1}$) and by the appearance of a new IR absorption band at 1700 $cm^{-1}$,
assigned for the stretching vibration $\nu$(C=O) of the isocyanurate
ring.[22] The properties of the model polycocyclotrimers are summarized in Table II.

The dynamic mechanical measurements revealed that the phase separation occurred in those polymers forming the soft segment (consisted
mainly of polyether chains) and hard segment (isocyanurate rings and
urethane groups) domains. The driving force for this phase separation
came from incompatibility (different solubility parameters) of segmental units. The $\alpha$-relaxation transition on the loss modulus E" (see
Figure 1) curve (maximum) is due to the microbrownian segmental motion
associated with the glass transition of the soft segment phase. This
Tg transition temperature ($\sim$ -30°C) is higher than Tg of the pure soft
phase (Tg of the polyoxytetramethylene is -79°C[23]) which indicates
that some phase mixing occurred.

In the case where POTMD of mol. wt. 2000 was used in the preparation of polycocyclotrimer, Tg value decreased to -67°C, which indicates a better phase separation. In general, a better phase separation is achieved when longer chain diol is used or when polycocyclotrimerization is performed in solution. Similar correlations were
reported previously for urethane block copolymers.[24,25]

The properties of polycocyclotrimers significantly changed with
the concentration of hard segments. The values of the storage modulus E' increased (and were less sensitive to temperature) with the
increase of the concentration of hard segments. (See Figure 2 and
Table II). The maximum of the loss modulus E" steadily decreased
and finally completely disappeared with the increase of the content
of the hard segment. Similarly, the maximum of the loss tangent (tan
$\delta$) shifted to the higher temperature under those conditions. All
mentioned changes in E', E" and tan $\delta$ are characteristic for the ad-
vancing crosslinking process (increase of the crosslink density) in
the hard segment.[26] At the same time, the relaxation transition
temperature of the crosslinked hard segment (as measured by DSC) increased with the increase of the molar ratio MDI/prepolymer. (Table
II). This phenomenon is consistent with the decrease of the segmental motion within hard segment due to the crosslinking reaction.[26]

The effect of the hard segment content on the stress-strain behavior of polycocyclotrimers was also investigated. The results are
summarized in Table II. It was found that with the increase of the
content of the hard segment, the tensile strength at ambient and
100°C increased and elongation decreased, and at the same time poly-

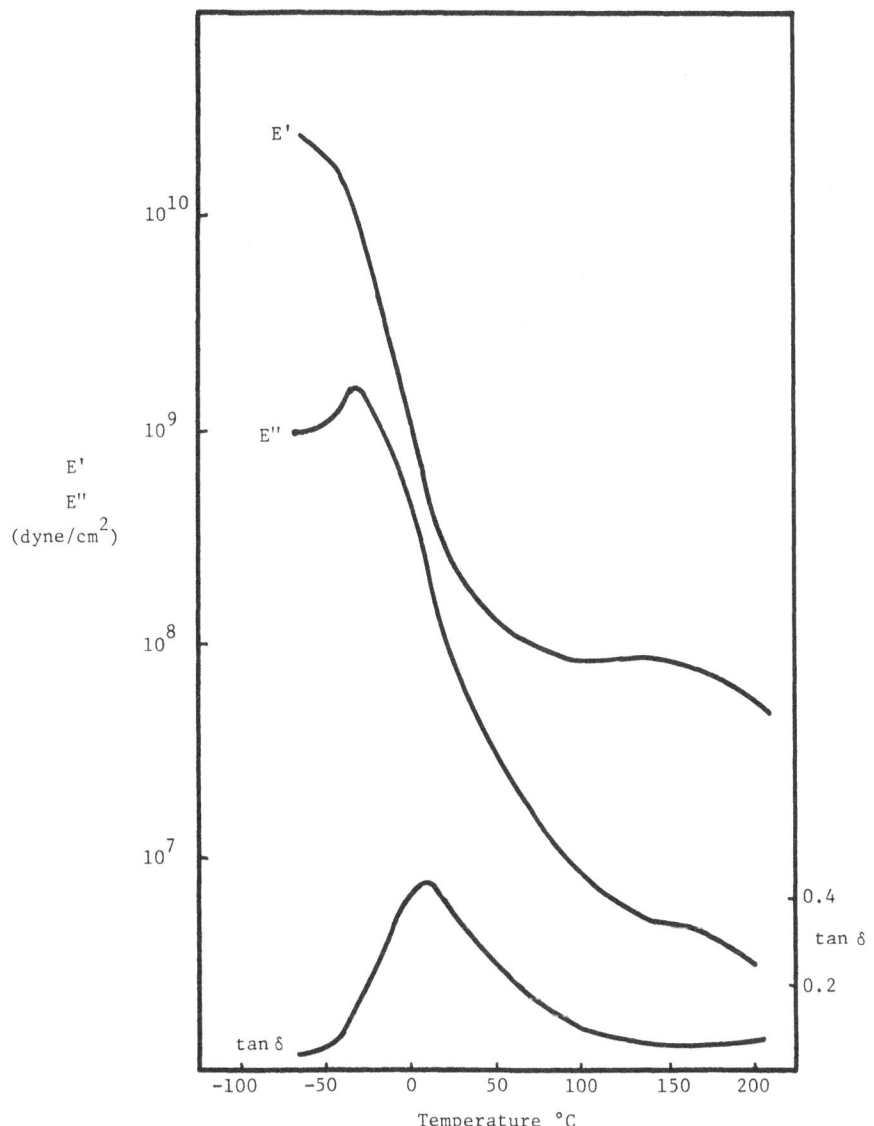

Figure 1.   Dependency of the storage (E'), loss (E") moduli and tan
            δ of polycyclotrimer (33.4% hard segments) on temperature.


cocyclotrimer changed from the elastomeric material to brittle, high
modulus plastic.  This change is associated with the inversion of
the continuous and dispersed phases which occur in polycocyclotrimer
at approximately 40-50% wt. of hard segment.  Similar behaviors were
described in other block copolymers containing soft and hard segments.[24]

Table II

The Properties of Polycocyclotrimers Prepared From MDI/MDI-POTMD Prepolymer

| Prepolymer | MDI/Prepolymer Mol. Ratio | Catalyst (% wt.) | Hard Segment Content (wt. %) | Tensile Strength MPa | | Elongation % | |
|---|---|---|---|---|---|---|---|
| | | | | 25°C | 100°C | 25°C | 100°C |
| MDI-POTMD 1000 | 0/1 | TDAPH (0.21) | 33.4 | 4.33 | 1.23 | 103 | 23 |
| MDI-POTMD 1000 | 1/1 | TDAPH (0.14) | 42.9 | 22.46 | 6.01 | 115 | 48 |
| MDI-POTMD 1000 | 2/1 | TDAPH (0.21) | 50.0 | 31.48 | 14.43 | 69 | 56 |
| MDI-POTMD 1000 | 4/1 | TDAPH (0.35) | 60.0 | 50.98 | 30.69 | 41 | 28 |
| MDI-POTMD 1000 | 6/1 | TDAPH (0.28) | 66.7 | 60.16 | 41.92 | 34 | 23 |
| MDI-POTMD 1000 | 8/1 | TDAPH (0.3) | 71.4 | 68.70 | 50.91 | 19 | 12 |
| MDI-POTMD 2000 | 2/1 | TDAPH (0.24) | 33.4 | 7.62 | 3.68 | 34 | 15 |

Table II (Continued)

The Properties of Polycocyclotrimers Prepared From MDI/POTMD Prepolymer

| Prepolymer | MDI/Prepolymer Mol. Ratio | Catalyst | Hard Segment Content (wt. %) | E'' Viscoelastic Transition Tg (°C) | Endotherm (°C) (DSC) |
|---|---|---|---|---|---|
| MDI-POTMD 1000 | 0/1 | TDAPH | 33.4 | -32 | 196 |
| MDI-POTMD 1000 | 1/1 | TDAPH | 42.9 | -22 | 195.2 |
| MDI-POTMD 1000 | 2/1 | TDAPH | 50.0 | -30 | 199.5 |
| MDI-POTMD 1000 | 4/1 | TDAPH | 60.0 | - | 206.6 |
| MDI-POTMD 1000 | 5/1 | TDAPH | 66.7 | - | 220.5 |
| MDI-POTMD 1000 | 8/1 | TDAPH | 71.4 | - | - |
| MDI-POTMD 2000 | 2/1 | TDAPH | 33.4 | -67 | 204.9 |

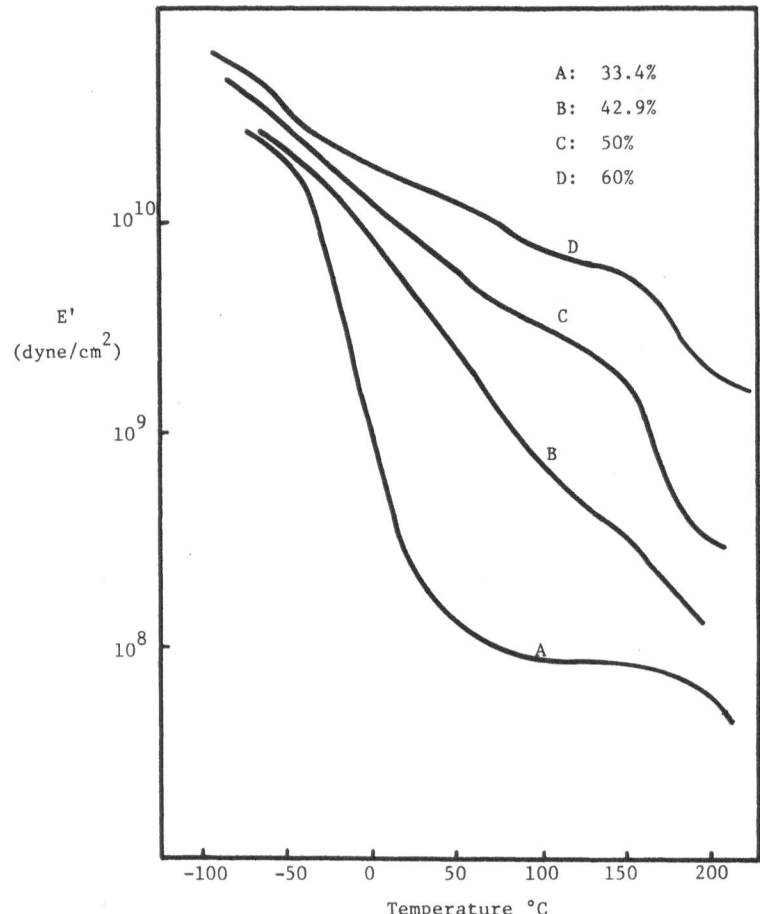

Figure 2.   Temperature dependence of storage modulus (E') for MDI/
            MDI-POTMD 1000 polycocyclotrimers containing 33.4 (A),
            42.9 (B), 50 (C), 60 (D) wt. % of hard segments.


        In order to compare effects of urethane and isocyanurate set-
ments on storage modulus, a model urethane elastomer based on MDI,
POTMD 1000 and 1,4-butanediol (37.1% of hard segments) was prepared.
The dependency of storage moduli E' of polycocyclotrimer and polyurethane on temperature is presented in Figure 3.  As can be seen,
the effect of isocyanurate segments on modulus is more pronounced
than the comparable effect of urethane hard segments.

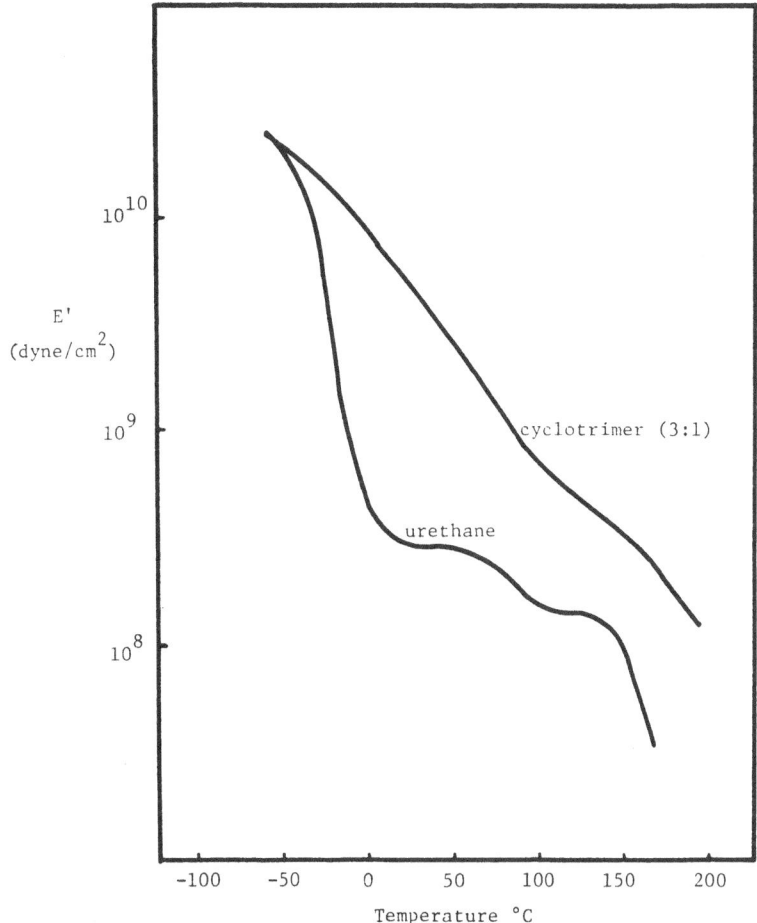

Figure 3.  Dependency of storage moduli (E') of polycocyclotrimer
           (42.9% hard segments) and polyurethane (MDI, POTMD 1000,
           1,4-butanediol, 37.1% hard segments) on temperature.

Preparation and Properties of Lower Index Urethane-Isocyanurate
Elastomers

     The urethane-isocyanurate elastomers with isocyanate indices
of 150, 200 and 250 were prepared from polymer polyol, 1,4-butane-
diol and an excess of isocyanate.  In this study the concentration
of the chain extender was kept at 15% weight based on the total
polyol component.  The mechanical properties of the resulting elastomers, such as tensile strength, flexural modulus, were determined
at different temperatures (-30°, 22°, 65° and 120°C).  The results
are summarized in Table III and Figures 4 and 5.  It was determined

Table III.  Properties of Urethane-Isocyanurate Elastomers

| Isocyanate Index | 150 | 200 | 250 |
|---|---|---|---|
| Tensile strength (psi): | | | |
| - 30°C | 8.319 | 8.3222 | 9.989 |
| 22°C | 4.785 | 5.513 | 6.579 |
| 65°C | 2.707 | 3.952 | 5.381 |
| 121°C | 1.140 | 2.429 | 4.076 |
| Tensile strength ratio: | | | |
| - 30/121 | 7.3 | 3.4 | 2.4 |
| Flexural modulus (psi): | | | |
| - 30°C | 146.220 | 227.581 | 228.368 |
| 22°C | 62.132 | 137.967 | 163.075 |
| 65°C | 20.987 | 81.632 | 85.211 |
| 121°C | 9.296 | 31.747 | 63.051 |
| Modulus ratio: | | | |
| - 30/65 | 6.9 | 2.78 | 2.68 |
| - 30/121 | 15.7 | 7.1 | 3.62 |
| Shore D hardness: | 72 | 77 | 81 |

that with the increase of the isocyanate index from 150 to 250, the tensile strength increased and the temperature tensile strength ratio decreased from 7.3 to 2.4.  This indicates that with the increase of the content of isocyanurate hard segments in elastomers, the tensile strength decreases in a slower manner with the increasing temperature.  A similar trend was also observed for the temperature dependence of the flexural modulus.  With the increase of the isocyanate index, flexural modulus also increased and the modulus ratio (-30°/65°) and (-30°/121°C) decreased.

The formation of isocyanurate crosslinks in the hard segment domains caused a significant increase in the tensile strength and flexural modulus and decreased their temperature dependency.  As a result of crosslinking of the hard segments, the elongation of the elastomer gradually decreased with the increase of the isocyanate index.

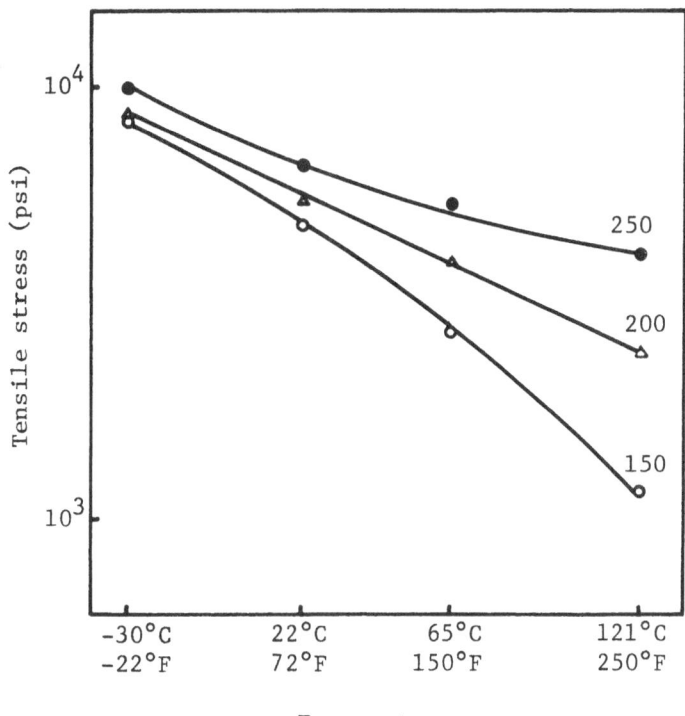

Figure 4.   Dependence of the tensile strength on temperature for
            urethane-isocyanurate elastomers.   Isocyanate indices:
            150, 200, 250.


SUMMARY

     The properties of urethane elastomers containing crosslinked hard
segments via isocyanurate groups were studied.   Various methods for
preparation of isocyanurate-urethane elastomers were described.   It
was found that isocyanurate-urethane copolymers had typical characteristics of phase separated block copolymers and that with the increase of the isocyanurate content, the tensile strength and flexural
modulus increased, and the temperature ratio of tensile strength and
flexural modulus and elongation decreased.   The reinforcement effect
of the isocyanurate crosslinks was more pronounced than the effect
of the urethane hard segments.   This phenomenon can be utilized in
the preparation of high modulus RIM elastomers.


Acknowledgements

     The financial support of the National Science Foundation (Grant

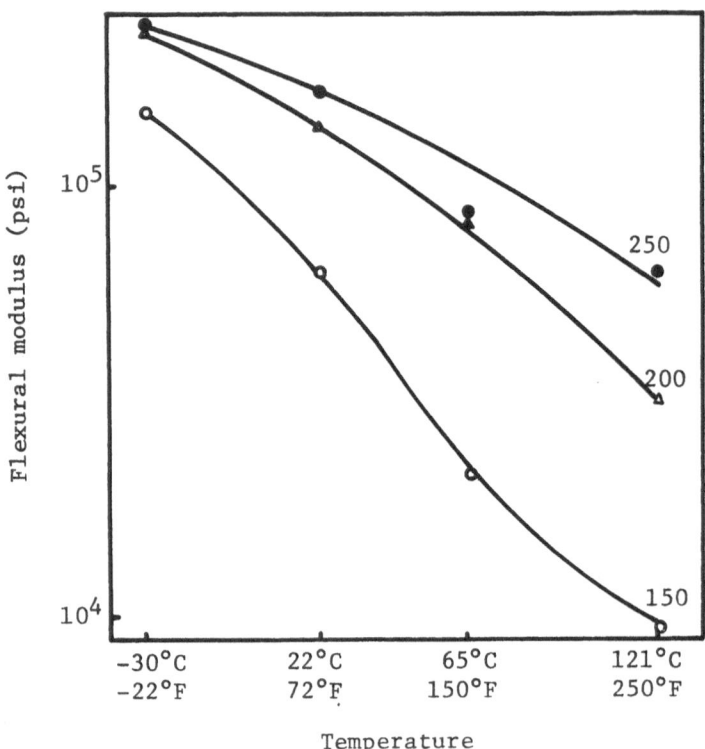

Figure 5.   Dependence of the flexural modulus on temperature for
            urethane-isocyanurate elastomers.   Isocyanate indices:
            150, 200, 250.


ENG 7825092) and of the Hexcel Corporation are gratefully acknowledged.


References

1.  Cooper, S. L. and Tobolsky, A. V., J. Appl. Polymer Sci., 10,
    1837 (1966).
2.  Bonart, R., Angew. Makromol. Chem. 58/59, 259 (1977).
3.  Blackwell, J. and Gardner, K. H., Polymer 20, 13 (1979).
4.  Huh, D. S. and Cooper, S. L., Polymer Eng. Sci., 11, 369 (1971).
5.  Lee, L. J., Rubber Chemistry and Technology, 153, 542 (1980).
6.  Srichatrapimuk, W. W. and Cooper, S. L., J. Macromol. Sci.,
    Phys. B15, 267 (1978).
7.  Friedman, I. R., Thomas, E. L., Lee, L. J., Macosco, C. W.,
    Polymer 21, 393 (1980).
8.  Assink, R. A. and Wilkes, G. L., J. Appl. Polym. Sci. 26,

3689 (1981).

9.  Shashoua, V. E., Sweeny, W, Tietz, R. F., J. Am. Chem. Soc. 82, 866 (1960).

10. Butler, G. B., Cornfield, G. C., J. Macromol. Sci. Chem., 5, 1889 (1971).

11. Butler, B. G., Cornfield, G. C., Aso, C., "Progress in Polymer Science," Jenkins, A. D., Ed., Pergammon Press, Oxford, 1975, 4, p. 112, 185.

12. Iwakura, Y., Uno, K., Ichikawa, K., J. Polym. Sci., A2, 3387 (1964).

13. Kordomenos, P., Kresta, J. E., Macromolecules, 14, 1434 (1981).

14. Sasaki, N., Yokoyama, T., Tanaka, T., J. Polym. Sci., A-1, 11, 248 (1970).

15. Kresta, J. E., Hsieh, K. H., Shen, S. C., Lin, I. S., SPE Technical Papers, 26, 462 (1980).

16. Saunders, J. H., Frisch, K. C., "Polyurethanes: Chemistry and Technology," Wiley-Interscience, New York, 1962, Part I.

17. Zdrahala, R. J. and Critchfield, T. E., ACS Organic Coat. Plast. Preprints, 42 (1980).

18. Gerkin, M., Lawler, L. F and Schwarz, E. G., J. Cellular Plastics, 51 (1979).

19. Kresta, J. E., Shen, C. S., Frisch, K. C., ACS Org. Coatings and Plastics Preprints, 36, 2, 674 (1976).

20. Kresta, J. E., Hsieh, K. H., Makromol. Chem., 179, 2779 (1978).

21. Kresta, J. E., Hsieh, K. H., Makromol. Chem., 180, 993 (1979).

22. David, D. J., Staley, H. B., "Analytical Chemistry of Polyurethanes," Wiley-Interscience, New York, 1969, p. 86.

23. Brandrur, J., Immergut, E. H., "Polymer Handbook," Wiley-Inter-Science, New York, 1967, p. III-80.

24. Van Bogart, J. W. C., Lilaonitkul, A., Cooper, S. L., "Multiphase Polymers"; Cooper, S. L., Estes, G. M., Eds., Advances in Chemistry Series, 176, ACS, Washington, DC, 1979, p. 3.

25. Seefried, C. G., Koleske, J. V., Critchfield, F. E., J. Appl. Polym. Sci., 19, 2493 (1975).

26. Murayama, T., "Dynamic Mechanical Analysis of Polymeric Materials," Elsevier, Amsterdam, 1978, p. 81-93.

POTENTIAL OF CROWN ETHER-ASSISTED ANIONIC POLYMERIZATION

OF STYRENE AND DIENES FOR COMMERCIAL PROCESSES

F. L. Cook, J. D. Muzzy, T. N. Montgomery,
R. M. Burton and K. E. Domeshek

Schools of Textile Engineering and Chemical Engineering
Georgia Institute of Technology
Atlanta, GA 30332

INTRODUCTION

Literally hundred of powerful cationic complexing agents have been prepared in the past 25 years, with the synthetic multidentate macrocycles receiving the greatest attention.[1] Macrocyclic oligomers of ethylene oxide, termed crown ethers, have warranted extensive investigation due to their extremely high binding constants for typical cations such as $Na^+$, $K^+$ and $Li^+$ compared to linear analogs.[2]

Although macrocyclic polyethers had been synthesized and characterized before 1967,[3] their ability to complex metal ions was not recognized. C. J. Pedersen of E. I. duPont, while synthesizing bis 2-(o-hydroxy phenoxy) ethyl ether from bis (2-chloroethyl) ether and the sodium salt of 2-(o-hydroxyphenoxy) tetrahydropyran containing residual catechol, isolated a small quantity (0.4% yield) of white, fibrous crystals.[4,5] The crystals were insoluble in neat methanol, but were readily solubilized by addition of sodium salt. Further investigations revealed the enhanced solubilization to be due to the sodium ion and not alkalinity. The by-product was identified via analysis as 2, 3, 11, 12- dibenzo-1, 4, 7, 10, 13, 16- hexaoxa-cyclooctadeca- 2, 11-diene.
Pedersen theorized that the sodium ion was apparently being held in the center of the macrocycle by electrostatic attraction between the positive charge of the sodium cation and the reinforced negative dipolar charge of the six oxygen atoms symmetrically arranged around it. Theoretical calculations by Frensdorff and Pedersen indicated that the hole diameter of the macrocyclic polyether dictated which specific cations would be most tightly bound, based on the Bohr's radius of the cation.[5,6] Pedersen logically concluded that presence

of the optimum cation during Williamson ether synthesis of the macrocycle should increase the percentage of cyclization versus oligomerization, with the cation "wrapping" around itself and consequently
bringing the ring-closing functionalities close together. Using the
so-called "template effect", Pedersen subsequently synthesized more
than 60 macrocyclic polyethers by condensation methods in yields
higher than allowed by strictly statistical arguments or by high-
dilution techniques.[5,7-9]

Pedersen reported a wide variety of stable alkali, alkaline
earth and other metal complexes with the macrocyclic polyethers in
non-polar aprotic as well as protic solvents. Many of the complexes
were isolated as sharp-melting crystals with phase transition temperatures far higher than those of the parent macrocycles. Molar
stoichiometries (polyether:cation) of 1:1, 2:1 and 3:2 were identified, with the former being the most prevalent.[8,9]

To avoid cumbersome IUPAC nomenclature, Pedersen also devised
an ad hoc system of names for the macrocyclic polyether.[10] Since
the molecular models resembled a crown, the epithet "crown" was applied to the class of macrocyclic polyethers, with the formal names
consisting of (in order): (1) the number and kind of hydrocarbon
rings, where applicable; (2) the total number of atoms in the polyether ring; (3) the class name (always "crown"); and (4) the number
of oxygen atoms contained in the polyether ring. Three examples of
unsubstituted crown ethers with the Pedersen nomenclature are shown
in Figure 1.

Synthetic Utilization of Crown Ethers

An explosion of publications have appeared on the synthetic
applications of crown ethers since the pioneering days of Pedersen
and co-workers.[1] Significant developments in the crown ether field
include high yield syntheses, enhanced purification procedures, rec-
ognition of the compounds as solid-liquid and liquid-liquid phase
transfer catalysts and incorporation of the macrocycles, either in
polymeric backbones or as pendant groups.[1,11-17] Many of the applications have utilized the "naked" character of the counterion to

12-CROWN-4          18-CROWN-6

15-CROWN-5

Figure 1.  Crown ether structures.

the complexed cation for organic syntheses.[14]  Since the coordination
sites of the cation are largely satisfied by the crown ether oxygens,
the corresponding anion is in essence solvent-separated from the positive charge.  As a result, the anion becomes a "harder", more highly
charged nucleophile than under normal circumstances.  The nucleophilicity is further enhanced if the crown ether is used as a solid-
liquid phase transfer catalyst to solublize the salt in a relatively
nonpolar medium.  The anion is then not only shielded from the carbon
by the crown ether but also is free of the tight solvation spheres
of water or other highly polar, protic solvent.

Facile syntheses of substituted crown ethers appeared several
years before those of their unsubstituted analogs.  As a result,
early studies of application of crown ethers to anionic polymerizations dealt with substituted derivatives.  Halasa and Cheng first
reported the polymerization of conjugated dienes using sodium or
potassium alkyls with tricyclohexyl- or bicyclohexyl-18 crown-6.[18]
Polydienes exhibiting 65-80% 1,2-configuration were isolated.  A
higher degree of conversion and a broader molecular weight distribution were obtained in systems utilizing crown ethers compared to
control systems.  The crown ether-sodium alkyl catalyst gave inherent viscosities of 2-10, whereas lithium alkyl gave low viscosities
($\leq$1).  Temperatures as high as 150°C were used without degradation
of the substituted crown ethers by the metal alkyls, whereas comparable linear ethers cleaved below 60°C.

Halasa and Cheng contributed the effect of crown ethers on conversion to stabilization of the growing allylic chain end by crown
ether chelation, preventing termination by disproportionation.[19]
The authors also reported the effect of crown ether addition on buta-
diene-styrene anionic copolymerization.

Alev et al prepared stable alkali metal solutions in low di-
electric solvents with dicyclohexyl-18-crown-6,[20] and initiated butadiene, isoprene and methyl methacrylate polymerizations with the systems.  Conversion of 70-100% were obtained, with considerably enhanced rates compared to heterogeneous metal-catalyzed systems.

In recent years an increasing number of researchers have recognized the benefits from incorporation of both substituted and unsubstituted crown ethers in anionic polymerization.  Slowkowski and
Penczek found that the dibenzo-18 crown-6/sodium acetate system gave
a 100-fold increase in the rate of polymerization of β-propiolac-
tone, allowing a lower temperature of reaction and decreased rates
of termination and transfer reaction.  Yamada used potassium acetate
with unsubstituted 18-crown-6- to increase molecular weight and
yields (to 99%) obtained in polyester formation from acrylic acid.[22]

A unique application of dicyclohexyl-18-crown-6 in the synthe-

sis of a triblock copolymer of poly(methyl methacrylate) and poly
(ethylene oxide) was reported by Suzuki et al.[23]  The initiator consisted of the crown ether complex of the disodium salt of poly(ethylene oxide).  The copolymer exhibited a higher syndiotactic triad
content in the PMMA block than polymers prepared by alternate methods.

Other reported polymerization systems utilizing crown ethers
include:  (1) oligomerization of ethylene oxide with 12-crown-4 or
18-crown-6 and sodium hydroxide in butanol;[24]  (2) oligomerization
of hexafluoropropene with cesium fluoride/18-crown-6 in THF[25a]; and
(3) bulk polymerization $\epsilon$-caprolactone and $\delta$-pyrollidone with sodium
hydroxide/18-crown-6.[25b]  Systems merging the common catalyst for
anionic polymerization of olefins and dienes, n-butyl lithium, with
unsubstituted crown ethers have not been extensively investigated.
In reference to the work of Halasa and Cheng,[19] unsubstituted crown
ethers are more powerful ligands than their substituted analogs.[1]
The enhanced binding constants are apparently due to the higher flexibility of the unsubstituted macrocycle, allowing the ligand to
"wrap" around the cation more effectively.  Although not as powerful
in binding ability as the related bicyclic cryptates[26] the unsubstituted crown ethers are much cheaper and more readily available.

## Crown Ether Potential in RIM and Solution Anionic Polymerization

The metal alkyls n-BuLi and sec-BuLi are now being widely ex-
ploited as polymerization initiators on a commercial scale, a trib-
ute to modern chemical engineering.  Most notably, styrene-butadi-
ene-styrene (SBS) and styrene-isoprene-styrene (SIS) terblock polymers are produced with lithium alkyls in solution polymerizations,
along with their SB and SI diblock analogs.[27]  Built around the
"living polymer" end concept first pursued by Szwarc,[28] the less
reactive dienes present the slow step in the processes.  The secondary metal alkyl is often used to optimize the rate of initiation
with the "harder" sec-butyl anion.  Crown ethers were predicted to
give a similar enhancement to the n-BuLi systems due to complexation of the lithium cation (and simultaneous formation of a "naked"
n-butyl anion) and prohibition of the typical associated form of
n-BuLi which is unreactive.[29]  The rate of propagation average molecular weight and molecular weight distributions of the polymer
systems were also predicted to be substantially altered on crown
ether complexation of the growing chain end.[18-19]

The same reasoning was applied to the potential of the n-BuLi/
crown ether system in reaction injection molding (RIM) of styrene.
RIM technology is growing rapidly since large parts can be formed
quickly in low pressure molds.  The process can result in significant energy and cost savings compared to conventional injection molding of preformed, high molecular weight polymers.  However, RIM has
not been used commercially for liquid vinyl systems, with existing

processes based on urethane technology.

Free radical polymerization has been examined as a reaction route for RIM processing of styrene by Kircher and co-workers.[30] Rapid free-radical polymerization leads to low molecular weight polymer unless crosslinking comonomers are added. However, the latter can lead to premature gelation and incomplete polymerization. In addition, any styrene polymerization method will result in a large heat of reaction and considerable shrinkage. A RIM process must compensate for these reaction characteristics.

Since termination chain growth polymerization normally occurs by chain transfer, anionic polymerization can be performed rapidly without sacrificing the molecular weight of the polymer. Oxygen, water and carbon dioxide, however, must be vigorously excluded due to their high reactivity with n-BuLi. Crown ethers were predicted to provide the acceleration needed to achieve the rapid reaction rates ( < one minute) desired in RIM.

The primary disadvantage of metal alkyl catalysts for anionic polymerization is the high reactivity of the reagent to impurities in the system. Alternate anions, such as "naked" acetate and cyanide formed by solubilizing the corresponding metal salts in organic solvents via crown ether addition, were considered as potential alternative candidates.[1]

Another candidate for crown ether-accelerated RIM processing is caprolactam.[31] In anionic ring opening polymerization of the monomer, a sodium metal dispersion in xylene is used to form the metal salt of the monomer:[27]

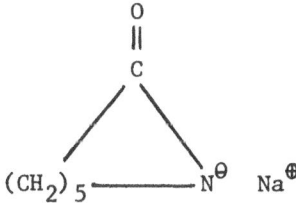

The metal salt then initiates the propagation reaction.

EXPERIMENTAL

The crown ethers utilized in the polymerization (12-crown-4, 15-crown-5 and 18-crown-6) were synthesized by the methods of Cook, Liotta et al.[11,13] The uniform synthetic approach included formation of the alkoxide of glycol ethers with a "template" metal base followed by displacement of halogen from a dichloride ether.[7] A

major portion of the water was removed from the hygroscopic crown
ethers by stirring the materials over Linde 4A molecular sieves for
10-12 hours. The macrocycles were then heated at 50°C under high
vacuum for several hours, sealed under vacuum and stored in a dessicator under nitrogen or vacuum until used.

Solvents and monomers were purified and dried by conventional
techniques.[32,34] The low shelf stability of the 1.6 M solution of
n-BuLi in hexane (Aldrich) dictated determination of the actual concentration of reagent every two weeks. The titration method of
Gilman and Cartledge was used to assess the n-BuLi concentration.[35]

Potassium acetate, potassium cyanide and sodium acetate (Fisher)
were dried at 125°C for 12 hours and utilized immediately upon removal from the oven. The salts were added to dry benzene containing
crown ether to form the stock solutions of "naked" anions. [1]H-NMR
spectra were obtained on a Varian Model EM-360A or T-60A spectrometer utilizing external TMS (2% in $CD_3Cl_2$) as a standard. A JEOL
PFT-100 Spectrometer was used to generate proton-decoupled [13]C-NMR
spectra by signal averaging (TMS standard). A Cannon-Fenske Vis-
cometer (ASTM 75) was used in all viscosity measurements, with the
solutions held at constant temperature ($\pm$ 0.2°F) in a controlled
aqueous bath. A Mettler Thermoanalyzer II was used to obtain DTA/
TGA scans. Molecular weight distributions were plotted either by
the Shell Oil Technical Center in Houston, Texas, or by use of an
LDC Constametric HPLC fitted with a $10^4\mu$ Styragel GPC column (Waters
Associates).

Polymerizations were conducted under an atmosphere of Linde
extra-dry nitrogen in a modified Labconco Model 5000 glove box. The
polymerization tubes were either especially constructed (dimensions
22 mm IDA, 25 mm ODA, 235 mm L), or were purchased from Ace Glass
Company (25 mm ODA, 10 mm thickness, 8 in. L). All systems were
sealed with Buna-n rubber septums (Texas Alkyls Co.) and metal crown
caps with 1/4 in. diameter opening to the septem (Aldrich). Ingredients were injected via Hamilton syringes fitted with Leur Locks and
6-inch, 20-gauge needles from Aldrich. After sealing, a multiarm
shaker from Burnell Corporation was used to agitate the samples in
ambient temperature polymerizations. A heated sand bath was utilized
for elevated temperature polymerizations.

In a typical n-BuLi catalyzed styrene or diene polymerization,
the solvent with dissolved crown ether was entered into the tube,
and the seal applied. Benzene was used as the solvent for styrene
polymerizations, and heptane for the diene polymerizations. Catalytic n-BuLi in hexane was then injected into the tube via hypodermic
syringe, and the tubes shaken until termination by methanol injection.

In the "naked" anion catalyst systems, the dried 18-crown-6 was added to benzene and the solution stirred 1.5 hours at ambient temperature with excess salt.  The solution was then separated from the remaining solid and the acetate anion concentration determined by $^1$H-NMR.  Based on the results of Dabdoub and Liotta,[36] the assumption was made that the solubility of potassium cyanide in benzene/18-crown-6 approximated that of potassium acetate.  The stock solution was added to styrene/benzene in the polymerization tubes at either ambient, 40°C or 75°C temperatures to initiate the chain growth.  Methanol was again used as a terminator.

In the attempted bulk polymerization of caprolactam, the dried and purified monomer was melted in a polymerization tube at >69°C.  A catalytic quantity of sodium dispersion in xylene was added to the molten monomer to generate the amide anion.  A pre-weighed sample of 18-crown-6 was then added to the system, the tube sealed and the mixture brought to the desired polymerization temperature range (150° - 260°C) by plunging the tube into a molten Wood's metal bath.

## RESULTS AND DISCUSSION

### Polymerizations With n-Butyl Lithium

The potential degradation of added crown ether by n-BuLi was an early concern in the research.  For example, the reaction path in Figure 2 is analogous to that observed in metal alkyl cleavage of linear ethers.[37]  The mechanism is likely ElcB or carbene-post elimination.  Such degradation products could lead to premature termination of the polymer chain or scavenging of the initiator, e.g., through Michael addition across the generated olefin.  Due to the high sensitivity and wide field of $^{13}$C-NMR compared to $^1$H-NMR, the former technique was chosen to qualitatively determine the degradative effect of n-BuLi on 15-crown-5 in benzene.  A $^{13}$C-NMR tube was partially filled with 4.97 mmoles of 15-crown-5 (4.14 M), 1 ml of benzene, TMS and 0.2 ml of benzene-d6 (for signal lock).  Similarly, a second tube was charged with n-BuLi (8.28 M) as the reagent.  The observed peak characteristics for the two tubes and the peak positions are detailed in Table I.

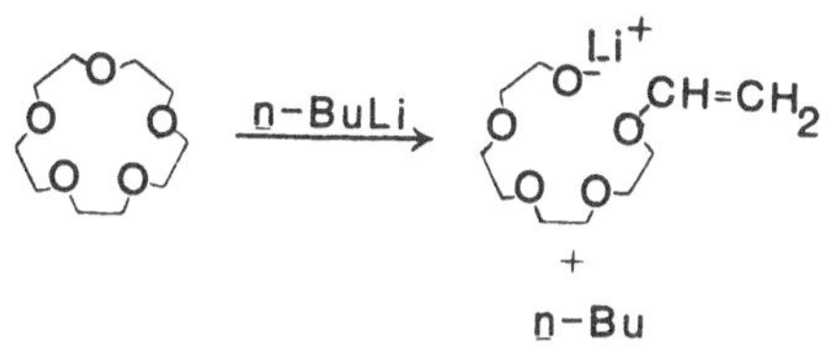

Figure 2.  Degradation of 15-crown-5 by n-BuLi

### Table I.  Reagent Peaks in $^{13}$C-NMR

| PEAK | POSITION | CHARACTERISTIC |
|------|----------|----------------|
| ∅ - H | 128.316 - 126.375 | TRIPLET |
| 15-CROWN-5 | 70.803 | SINGLET |
| n-BuLi/HEXANE | 35.004 - 14.191 | COMPLEX, 15 PEAKS |

The tube containing the crown ether was then injected with 6.62 mmoles of n-BuLi in 4.4 ml of hexane.  Final concentration in the tube was 0.93 M and 1.24 M for 15-crown-5 and n-BuLi, respectively.  Continuous $^{13}$C-NMR scans (2000 total) were accumulated over two hours at ambient temperature to monitor new peak formation and the effect on the parent peak.  Four new peaks appeared with time at the positions shown in Table II.  The 15-crown-5 peak at 70.803 ppm was still the prominent peak at the end of the two hour period, indicating that the degradation was slow and incomplete.  Subsequent polymerization results confirmed that the crown ethers were sufficiently stable in the kinetic lifetime of chain growth to dramatically affect the polymerizations.

A simplified schematic of the polymerization apparatus used for the styrene and diene systems with crown ether/n-BuLi is contained in Figure 3.  The first effect assessed was the change in percent conversion with time on crown ether addition.  To construct a baseline, an unaccelerated system consisting of 15 ml of benzene, 0.74 mm n-BuLi and sufficient styrene to generate 1.0 M and 2.0 M concentrations were studied.  The results are exhibited in Figure 4. As expected from traditional kinetic expressions,[29] the overall rate decreased as a function of decreasing styrene concentration.

With 43 mmoles of styrene in the above conventional system (2 M), 100% conversion was obtained at ambient temperature after 20

### Table II.  Results of $^{13}$C-NMR Study

| PEAK | POSITION (PPM) | CHARACTERISTIC |
|------|----------------|----------------|
| CROWN-CH$_2$- | 70.803 | SINGLET |
| NEW PEAKS | | |
| 1 | 151.430 | W.SINGLET |
| 2 | 85.849 | W.SINGLET |
| 3 | 69.469 | W.SHOULDER ON CROWN PEAK |
| 4 | 67.406 | W.SINGLET |

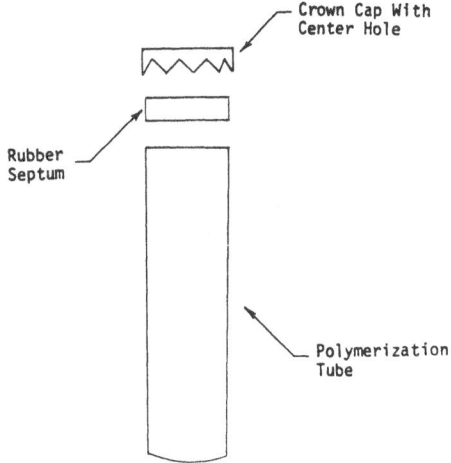

Figure 3.   Polymerization apparatus.

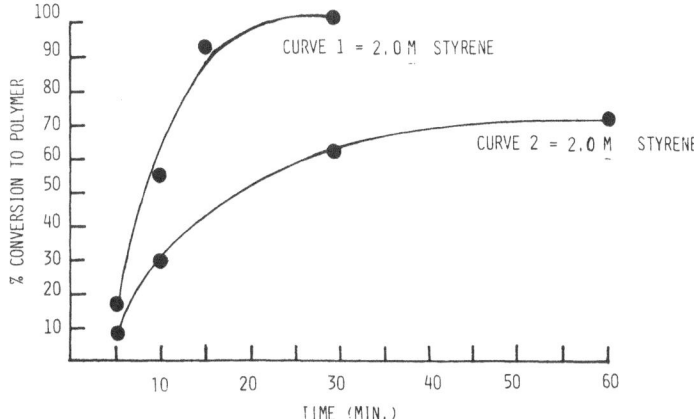

Figure 4.   Overall rate study for styrene polymerization systems
            containing no crown ether.


minutes.  Adding 0.189 mmoles of 15-crown-5 to the same system (25
molar percent of catalyst) gave complete conversion in less than 10
seconds.  Ten seconds was the least measurable time from n-BuLi injection to methanol injection and termination.  The rate of reaction
as a function of 15-crown-5 concentration is shown in Table III.

     Direct molar substitution of 12-crown-4 for 15-crown-5 in the
polymerization system gave a reduced rate, with 100% conversion requiring 2.1 minutes.  Percent conversion as a function of time for
the 12-crown-4 system is detailed in Figure 5.  Although the rate

Table III.   Crown Systems in Polymerization Styrene

| TUBE | BENZENE (ML) | STYRENE (MMOLES) | 15-CROWN-5 (MMOLES) | N-BuLi (MMOLES) | TIME OF REACTION (SEC.) | CONVERSION OF POLYMER (%) |
|------|--------------|------------------|---------------------|-----------------|-------------------------|---------------------------|
| 1    | 15           | 43               | 0.074               | 0.74            | 10                      | 92.5                      |
| 2    | 15           | 43               | 0.189               | 0.74            | 10                      | 100                       |
| 3    | 15           | 43               | 0.063               | 0.74            | 15                      | 100                       |
| 4    | 15           | 43               | 0.179               | 0.74            | 30                      | 100                       |
| 5    | 15           | 43               | 0.00                | 0.74            | 1200                    | 100                       |

was slower than with 15-crown-5, it was still ten times faster than that of the standard (crown-free) polymerization. The indirect evidence indicated that 15-crown-5 complexes lithium more effectively than 12-crown-4, contrary to the theoretical calculations of Frensdorff[6] but recently supported by evidence of other researchers.

The effect of crown ether concentration and more specifically the crown ether:butyl lithium molar ratio on the viscosity average molecular weight of polystyrene was also of interest. First, a series of polymerizations were conducted by varying the n-BuLi concentration, and the inherent viscosity ($\eta_{inh}$) was determined for the resulting polymers. The tube contents for the series are relayed in Table IV. The same series was then repeated with a set addition of 0.19 mmoles of 15-crown-5 to each polymerization tube.

The results are exhibited in Figure 6. Although the two curves followed the same pattern, the curve for the crown ether system was uniformly shifted to a higher $\eta_{inh}$. The gap between the two curves was greatest at the lowest n-BuLi concentration and gradually narrowed until near-convergence at high n-BuLi concentration. In essence, the gap was widest at the highest crown ether:n-BuLi molar ratio (0.27:1) corresponding to the highest percentage of growing

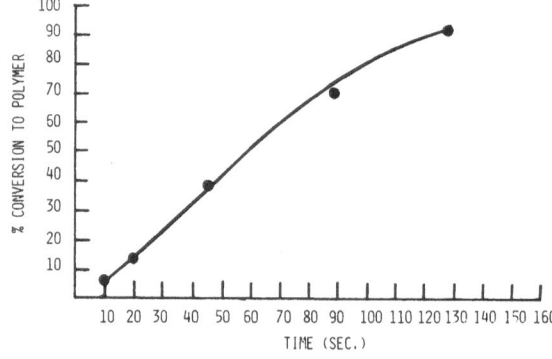

Figure 5.   Rate study for polymerization of styrene using 12-crown-4 (0.01 M).

Table IV.   Systems for Assessing Effect of n-BuLi Concentration on Styrene Polymerization.

| TUBE | BENZENE (ML) | STYRENE (MMOLES) | N-BuLi (MMOLES) | N-BuLi CONCENTRATION (M) |
|------|--------------|------------------|-----------------|--------------------------|
| 1 | 15 | 43 | .70 | .033 |
| 2 | 15 | 43 | .98 | .045 |
| 3 | 15 | 43 | 1.96 | .082 |
| 4 | 15 | 43 | 2.94 | .113 |

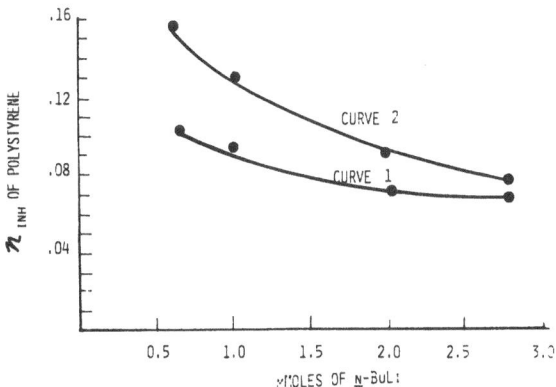

Figure 6.   Effect of n-BuLi concentration on inherent viscosity of polystyrene prepared with no crown (Curve 1) and with 15-crown-5 (Curve 2).

chain ends with the lithium cation complexed. At the convergent point, the ratio was at its lowest level, (0.06:1), with only 6% of the theoretical carbanion chain ends in the "naked" mode. The crown ether was obviously dramatically affecting the degree of polymerization of the complexed chain ends.

The molar ratio of crown ether:n-BuLi was then systematically altered by gradually increasing the concentration of crown ether while maintaining n-BuLi concentration. The results for both 15-crown-5 and 12-crown-4 are shown in Figure 7. The viscosity average molecular weights of the synthesized polystyrenes peaked at a common molar ratio of 0.5:1, and decreased again in near-symmetrical fashion as the ratio approached 1:1. Significantly, 15-crown-5 gave higher $\overline{M}_v$'s than did 12-crown-4 at all molar ratios (3.65 x $10^4$ grams vs. 3.14 x $10^4$ grams at the peaks). The 15-crown-5 and 12-crown-4 systems gave maximum $\overline{M}_v$'s 1.8X and 1.6X, respectively, higher than identical systems without crown ether.

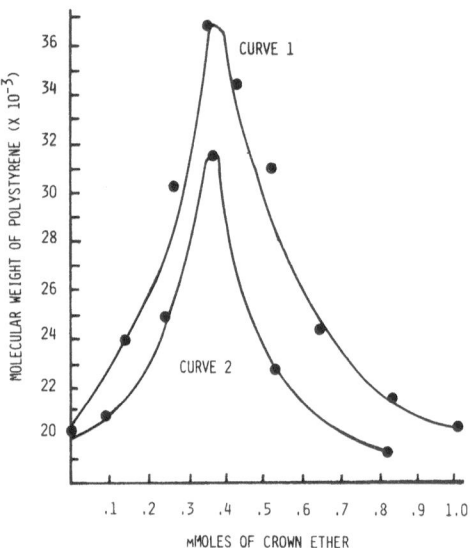

Figure 7.   Effects of 15-crown-5 (Curve 1) and 12-crown-4 (Curve 2)
            on molecular weight of polystyrene under identical
            polymerization conditions.


        Gel permeation chromatography (GPC) was used to further eluci-
date the effects of adding the two crown ethers to the styrene polymerization system.   The GPC traces are exhibited in Figure 8.   Polydispersity indices were determined as 2.0 (15-crown-5), 1.2 (12-crown-
4) and 1.2 (no crown), indicating broadening.   Number average molecular weights ($\overline{M}_n$) for the three systems as determined at the scan peaks
were 31, 20 and 9.6 x $10^3$, respectively, following a similar pattern
as the $\overline{M}_v$'s.

        The overall rate, average molecular weight and polydispersity
index data for the synthesized polystyrenes can be directly attributed to the population of crown ether-complexed species.   Kinetically, the effect of the crown ether can be pictured as a modification
of the scheme of Worsfold and Bywater for styrene polymerization in
benzene with n-BuLi.[29]   At moderate concentrations, the n-BuLi (BL)
is associated:

$$(BL)_m \underset{\longleftarrow}{\overset{K_1}{\longrightarrow}} mBL \tag{1}$$

where $K_1$ is the equilibrium constant and m is the degree of association (normally 6).   At equilibrium the associated species, which do
not react with styrene, are formed.   Consequently, fewer molecules of
active initiator are available for polymerization.   The initiation
reaction is:

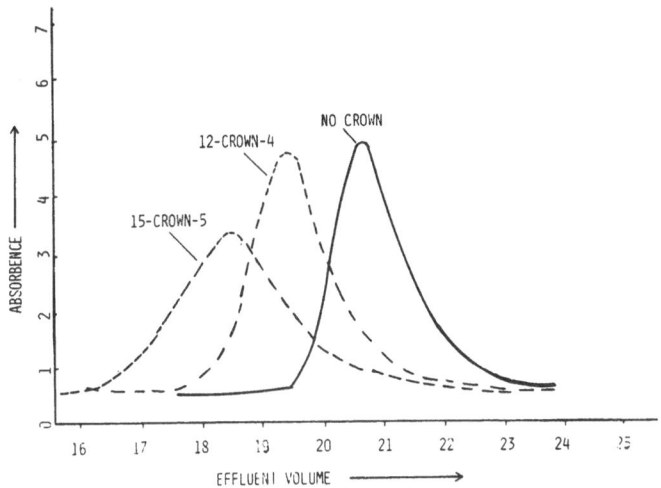

Figure 8.   GPC traces of reaction mixtures

$$BL + S \xrightarrow{k_2} BSL \tag{2}$$

where $\underline{S}$ is styrene and $k_2$ the rate constant.  Propagation is independent of the degree of polymerization, $\underline{x}$:

$$BS_xL + S \xrightarrow{k_3} BS_{x+1}L \tag{3}$$

The growing polystyryl chain ends can associate:

$$(BS_xL)_2 \underset{}{\overset{K_4}{\rightleftharpoons}} 2BS_xL \tag{4}$$

and the associated bipolymer is not reactive with styrene.  Polymer growth usually terminates by chain transfer:

$$BS_xL + TR \xrightarrow{k_5} BS_xT + LR \tag{5}$$

where TR is a transfer agent such as water, oxygen or carbon dioxide.

With crown ether, $\underline{CE}$, present, the first reaction is to form a complex with $\underline{n}$-BuLi:

$$BL + CE \underset{}{\overset{K_6}{\rightleftharpoons}} BL \cdot CE \tag{6}$$

Due to high binding constant of the crown ether for the lithium cation and the favorable polar interactions, complex formation is preferred over association of n-BuLi (Reaction 1). Consequently, a higher concentration of active n-BuLi is available for polymerization. The initiation reaction is:

$$BL \cdot CE + S \xrightarrow{k_7} BSL \cdot CE \qquad (7)$$

followed by the propagation reaction:

$$BS_xL \cdot CE + S \xrightarrow{k_8} BS_{x+1}L \cdot CE \qquad (8)$$

Effective complexation of the lithium cation shields the n-butyl and chain anions from the metal's positive charge, giving "naked", harder nucleophiles that potentially propagate at a faster rate. Kinetically, the rate constants $k_7$ and $k_8$ should be significantly greater than the respective rate constants without crown ether, $k_2$ and $k_3$. Complexation also contributes to reduced n-BuLi association (larger $K_1$), yielding higher active initiator concentration and increasing the reaction rate.

The polymer may be terminated by chain transfer:

$$BS_xL \cdot CE + TR \xrightarrow{k_{10}} BS_xT + LR + CE \qquad (10)$$

Conservatively, the transfer reaction should take place at the same rate as in Reaction 5, with the crown ether having little or no effect on the rate of termination. Halasa and Cheng claimed that the effect of crown ethers on conversion was explainable in terms of stabilization of the growing allylic anion by crown ether chelation.[19] The chelation was pictured as stabilizing the growing end of the polymer chain and preventing termination by disproportionation. However, as detailed in the above kinetic scheme, the rate, $\overline{M}_v$ and GPC date can be adequately explained by the "naked" anion theory and the decrease in association of n-BuLi. In addition, termination of the highly active chain end by chain transfer is more likely than disproportionation.

The broadening of the molecular weight distributions on crown ether addition was also unexpected in view of earlier systems with added tetrahydrofuran (THF). The addition has been studied by Szwarc and others as a means of increasing the proportion of solvent-separated ion pairs in the anionic polymerization of styrene.[38] The propagation rate constant $k_-$ for the free anion was shown to be much faster than the ion pair rate constant $k_+$. Richards noted that the presence of species with vastly different propagation rate constants should result in polymers possessing broad distributions.[39] However, data for the systems containing THF show that the polydispersity indices are close to the predicted Poisson distribution, analogous to the conventional system without THF. Richard ascribed the anomaly to extremely rapid rates of exchange between the contact ion pairs, solvent separated ion pairs and free ions in the system. As a result, the growing chain end was theorized to experience all three modes many times during its growth cycle. The rapid exchange promotes an averaging-out effect on the chain length, and hence results in the traditional Poisson distribution for anionic-polymerized systems. Molecular weight broadening has been induced by the use of added salts,[40] and the rates of interconversion have been estimated to be extremely rapid.[41]

Exchange of the crown ether between the complexed chain end and active species in the styrene system is denoted by:

$$BS_xL \cdot CE + BL \overset{K_7}{\rightleftharpoons} BS_xL + BL \cdot CE \qquad (11)$$

$$BS_xL \cdot CE + BS_x'L \overset{K_8}{\rightleftharpoons} BS_xL + BS_x'L \cdot CE \qquad (12)$$

where $BS_x'L$ denotes an uncomplexed growing chain end. Since the broadening of the molecular weight distributions in Figure 8 was not extreme as evidenced by the polydispersity indices, a slower rate of exchange of the crown ether compared to the analogous THF is likely. Halasa and Cheng had also observed a broadening of diene molecular weight distribution upon substituted crown ether addition.[19] The broadening attained with crown ether offers possible advantages in final polymer mechanical properties over the conventional or THF systems.

Determination of accurate rate constants was not attempted with the initial data due to possible interference with exothermal acceleration. The polymerizations were performed in the heavy-walled (10 mm) sealed tubes. Theoretical calculations indicated that if 90% conversion was attained in 10 seconds, less than 10% of the generated heat would escape through the walls. Reducing the styrene concentration was proposed as a route to reduced thermal acceleration. However, decreasing the styrene concentration while maintaining the optimum styrene:n-BuLi:15-crown-5 ratio led to increasing molecular weight. For example, decreasing [S] from 0.8 $\underline{M}$ to 0.2 $\underline{M}$ at a fixed 70:$\underline{1}$:0.5 molar ratio of styrene:n-BuLi:15-crown-5 gave an increase in $M_v$ from 17,000 to 79,000. The data also exhibited more scatter at the lower concentrations of reagents. The $M_v$ trend in conjunction with the scatter suggested that intractable impurities were scavenging a portion of the n-BuLi. In fact, 100% conversion could not be obtained below a 0.2 $\underline{M}$ styrene concentration in the scheme. At the lower concentration of styrene and hence $\underline{n}$-BuLi, the initiator was apparently not in sufficient mass to completely eliminate the impurities. The latter then prematurely terminated chain ends and prevented growth to high $M_v$.

As shown in Table III, decreasing the crown ether concentration by lowering the optimum ratio slowed the overall rate of reaction, but only slightly (Tubes 1 and 3 versus Tube 2). In addition, alteration of the optimum rates dictated a penalty in decreased $M_v$ (Figure 7).

The conversion to polymer versus time under dilute conditions has been reinvestigated using thinner-walled glass tubes (3 mm). At 0.2 $\underline{M}$ styrene, the maximum possible temperature rise was calculated as 10°C assuming adiabatic conditions. A period of 250 seconds was required to reach 100% conversion under the dilute conditions at the optimum ratio, with the plot of percent conversion versus time deviating slightly from linearity and exhibiting some scatter. A short induction period of approximately five seconds was also apparent. Part of the induction period was attributed to dispersal of the $\underline{n}$-BuLi on injection, and part to the rates of Reactions 6, 7 and 10. The bulk of the conversion is due to propagation (Reaction 8). Ignoring the slower, unaccelerated propagation Reaction 3 and crown ether exchange, the kinetic expression for styrene depletion is:

$$\frac{-d[S]}{dt} = k_8 \; [BS_x L \cdot CE] \; [S] \tag{13}$$

As an approximation, the concentration of crown ether initially charged was substituted for the concentration of growing polymer:

$$\frac{-d[S]}{dt} = k_{obs} \; [CE] \; [S] \tag{14}$$

and $k_{obs}$ was evaluated from the percent conversion versus time data at 0.2 $\underline{M}$ [S]. In the 10 - 60% conversion range, the observed rate constant was 10 1/mole-sec. The value is close to the estimated propagation rate constant for the system with no crown ether.[42] The overall rate of conversion, however, is dramatically faster for the systems containing crown ether (Figure 5 and Table III). Unlike the conventional system, the rate-controlling step in the crown ether-accelerated polymerization is apparently dictated by Reaction 8 and not the disassociation Reaction 1 or initiation Reaction 7.

Polymerization of isoprene in heptane with the $\underline{n}$-BuLi/unsubstituted crown ether complex as initiator followed the general pattern of the styrene research. The systems investigated with both 15-crown-5 and 12-crown-4 addition are detailed in Table V. Overall rate comparisons of accelerated and conventional polymerizations are

Table V.  Crown Systems in Isoprene Polymerization

| TUBE | HEPTANE (ML) | ISOPRENE (MMOLES) | 15-CROWN-5 OR 12-CROWN-4 (MMOLES) | N-BuLi (MMOLES) | CONC. OF N-BuLi (M) | CONC. OF CROWN (M) |
|------|------|------|------|------|------|------|
| 1 | 15 | 50 | 0 | 1.49 | 0.075 | 0 |
| 2 | 15 | 50 | 0.048 | 1.49 | 0.075 | 0.0023 |
| 3 | 15 | 50 | 0.084 | 1.49 | 0.075 | 0.0040 |
| 4 | 15 | 50 | 0.127 | 1.49 | 0.075 | 0.0060 |
| 5 | 15 | 50 | 0.274 | 1.49 | 0.075 | 0.0130 |
| 6 | 15 | 50 | 0.398 | 1.49 | 0.075 | 0.0189 |

plotted in Figure 9.  As with styrene, both crown ethers gave considerably faster overall rates of conversion than the system without
the macrocycle.  Again, 15-crown-5 was more effective than 12-crown-4
in rate enhancement, but the difference was not as dramatic as in the
styrene polymerizations.

$\overline{M_v}$ plots as a function of crown ether:$\underline{n}$-BuLi molar ratio for the
isoprene polymerizations are shown in Figure 10.  Although following
the same pattern of increasing $\overline{M_v}$ at lower molar ratios, the data was
limited to 0.17:1 and 0.32:1 for 15-crown-5 and 12-crown-4, respectively.  At higher molar ratios, injections of $\underline{n}$-BuLi resulted in
a rapid (less than 10% conversion to polymer) precipitation of a
yellow-brown solid.  Although difficult to analyze, the precipitate
was theorized to be a crown ether-lithium salt complex occluding from
the supersaturated, nonpolar heptane solvent.  An analogous system is
the crystallization of the acetonitrile/18-crown-6 charge transfer
complex from solvents above the critical concentration of the macrocycle.[11]  In the polymerization system, the anion of the complex was
likely polyisoprenyl or butyl in nature.

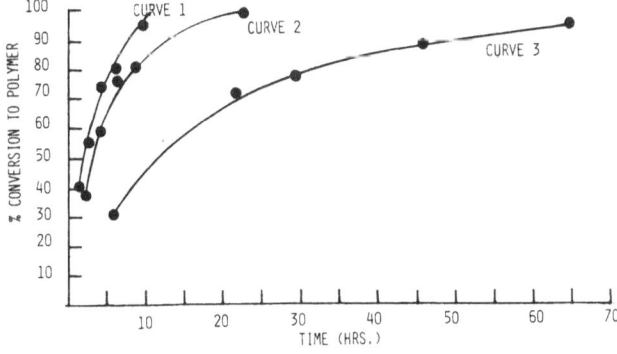

Figure 9.  Rate comparison for isoprene polymerization systems containing 0.114 mmoles of 15-crown-5 (Curve 1), 0.114 mmoles
of 12-crown-4 and no crown (Curve 3).

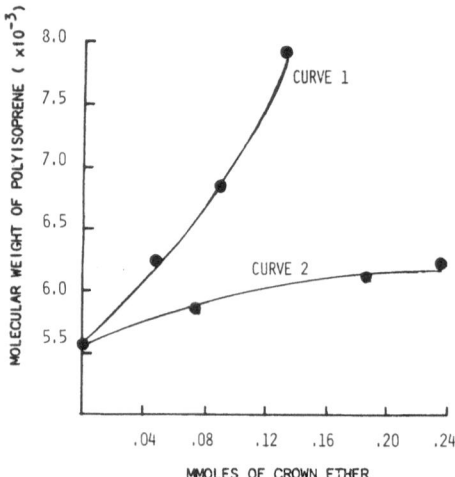

Figure 10.   Comparison of effects of 15-crown-5 (Curve 1) and 12-crown-4 (Curve 2) on molecular weight polyisoprene under identical polymerization conditions.

Attempts to translate the crown ether-accelerated isoprene polymerization system to butadiene revolved around Union Carbide's CP commercial grade of the monomer.  In commercial anionic SB and SBS processes, the butadiene is produced in-house under vigorous purity controls and routed directly into the reactor.  By contrast, the CP grade of material contains the impurity concentrations detailed in Table VI.  Not having access to high purity butadiene, the authors attempted to remove the impurities from the CP grade monomer by combining in series recommended techniques.

For example, in one set of polymerizations the gaseous butadiene was passed through a tube series containing (in order) activated alumina and silica gel.  The monomer was then condensed in a polymerization tube containing excess n-BuLi as an impurity scavenger at dry ice temperatures.  The tube temperature was then allowed to rise slowly, and the butadiene was distilled into a second polymerization tube.  The other system components (crown ether dissolved in heptane,

Table VI.   Union Carbide CP Grade Butadiene

|  | PPM |
| --- | --- |
| ACETYLENES | 600 |
| ISOBUTYLENE | 500 |
| 1-BUTENE | 2000 |
| t-2-BUTENE | 2000 |
| BUTADIENE DIMER | 50 |
| 4-VINYL-5-CYCLOHEXANOL | 6000 |
| DI-t-BUTYL CATECHOL | 120 |

followed by $\underline{n}$-BuLi/hexane) were then injected into the tube in the prescribed manner.

Despite the extensive purification attempts, inherent viscosities of the synthesized polybutadienes prepared in the heptane-$\underline{n}$-BuLi systems fell in the low 0.038 - 0.052 dL/gm range at 30 - 60% conversion with or without crown ether addition. The tenacious impurities in the CP grade of butadiene were apparently not completely removed by the combined techniques, leading to premature termination, low yields and low molecular weights. For example, acetylenes have been reported to terminate anionic butadiene polymerizations at levels as low as 2 ppm,[43] compared to an initial level of 600 ppm in the CP monomer grade (Table VI).

## Polymerizations Initiated by "Naked" Acetate and Cyanide

"Naked" acetate and cyanide anions were investigated as alternative initiators to the butyl anion in the polymerization of styrene.[14,44] The former are less reactive than the latter with such common impurities as $H_2O$, $CO_2$ and $O_2$.

[1]H-NMR studies of the 18-crown-6/benzene solution stirred over excess potassium acetate at 25°C revealed that the crown ether:salt molar ratio remained at the reported 1:1[44] only in the 0.15 - 0.20 $\underline{M}$ macrocycle concentration range. Significant deviation from a 1:1 molar ratio was observed at crown ether concentrations both higher and lower than 0.15 $\underline{M}$. Solubility of complexed potassium cyanide was assumed to parallel that of potassium acetate.[36] Stock benzene solutions were prepared of known 18-crown-6 (and hence "naked" acetate or cyanide) concentrations by stirring for 1.5 hours over the respective solid salts at room temperature.

In a typical polymerization of styrene with the "naked" anions, a tube was charged with 5 ml (43.3 mmoles) of styrene under vigorous exclusion of water and air. Sufficient stock solution (17.5 ml of 0.005 $\underline{M}$ 18-crown-6 and 0.067 $\underline{M}$ potassium acetate by NMR) was then added to generate a final anion concentration of 0.052 $\underline{M}$ (1.17 mmoles of "naked" acetate). Final concentration of crown ether was 0.004 $\underline{M}$ (0.087 mmoles), and monomer concentration was 1.9 $\underline{M}$. The sealed tubes were then placed in a shaker at ambient temperature or in a static sand heating bath at 40°C or 75°C for 12 hours. Methanol was again used as a terminator.

"Naked" acetate gave optimum polymerization results at 40°C (6.4% conversion, $\eta_{inh}$ 2.43 dL/gm). "Naked" cyanide gave optimum results at 75°C (6.0% conversion, $\eta_{inh}$ 2.46 dL/gm). The inherent viscosities were the highest observed for any polystyrene samples synthesized in the crown ether research, including those initiated with $\underline{n}$-BuLi. For example, the $\eta_{inh}$ for the peak sample in Figure 7 with the 15-crown-

5/n-BuLi catalyst was 0.195 dL/gm.  Blank runs without 18-crown-6
gave no polymer under the same conditions.  The low conversions and
high degrees of polymerization were likely due to impurity scavenging
by the solubilized anion, leaving few "naked" anions to generate chain
growth.  The starting crown ether:salt molar ratio was only 0.07:1.
The few chain ends formed by the residual "naked" anions grew to high
molecular weights before termination, since the bulk of the impurities
had apparently been scavenged in the early stages by the catalyst.

## Crown Ether Addition to Caprolactam Polymerization

The conventional bulk anionic polymerization of caprolactam is
initiated by the sodium salt of the monomer.[27]  The salt is formed by
addition of a sodium metal in xylene to the molten monomer at a 0.04 –
0.08 mole % level.  In conventional runs, a temperature of 255 – 265°C
gave 100% conversion to polymer in 5-10 minutes.

For crown ether-assisted polymerizations, the molar ratios of
caprolactam:sodium metal:18-crown-6 was held at 166:1:1.5 and the
polymerization temperature was varied through the series shown in
Table VII.

Brown discoloration of the reaction mixture occurred upon crown
ether addition in each case.  The discoloration apparently resulted
from crown ether degradation, as identical systems without the macrocycle did not develop color.  A hot $H_2O$ extraction revealed that no
polymer was formed in the Table VII systems.  The conventional systems gave 100% yield.

Halasa and Cheng have reported a limit of 150°C at which bi- or
tri-cyclohexyl-18-crown-6 can be used in solution anionic polymerizations.  Differential thermal analysis/thermogravimetric analysis
(DTA/TGA) of unsubstituted 18-crown-6 performed under air atmosphere.
The plots are detailed in Figure 11.  The compound begins to thermally
degrade at ∿ 120°C, shows substantial weight loss at 180°C and is
fully decomposed at 200°C.

In the purification of 18-crown-6 by vacuum distillation (130°C,
0.2 mm), several groups have reported explosions resulting from vent-

Table VII.  Crown Systems in Polymerization of Caprolactam

| TUBE | Na (G) | 18-CROWN-6 | REACTION TEMPERATURE (°C) | REACTION TIME (HR.) |
|------|--------|------------|---------------------------|---------------------|
| 1 | .0190 | 1.18 | 200 | 4 |
| 2 | .0195 | 1.24 | 180 | 4 |
| 3 | .0195 | 1.14 | 150 | 4 |
| 4 | .0201 | 1.21 | 110 | 48 |

ing hot stills to air.[45]  The explosions were attributed to autooxi-
dation of hot dioxane vapors, with the stable 6-member heterocycle
formed by thermal ring contraction of 18-crown-6.   The probable mechanism of thermal degradation and rearrangement is shown in Reaction
15.

(15)

Reaction 15 does not account, however, for color formation nor
for failure to isolate polymer from the systems containing crown
ether.  The lactam anion was apparently reacting with the crown ether
or one of its thermal degradation products to deprive the system of
initiator.  Since the sodium salt is analogous to n-BuLi, a degradation reaction similar to that detailed in Figure 2 may have occurred.
Chemical attack on the crown ether was also supported by the unsuc-

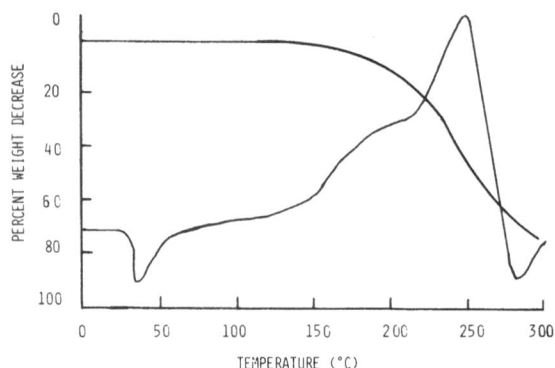

Figure 11.   DTA/TGA of 18-crown-6 under air.


cessful polymerization at a holding temperature of 110°C, well below
the thermal degradation temperature of 18-crown-6 (Figure 11).


CONCLUSIONS

[13]C-NMR studies revealed that unsubstituted 15-crown-5 is sufficiently stable in the presence of n-BuLi in benzene at room temperature to render beneficial results in a time frame consistent with
chain growth kinetics.[27]  Although the crown ether peak was still
strong and prominent in the spectra after two hours, several minor
peaks appeared in the spectra with time, indicating some cleavage.
The cleavage products were theorized to be alkenes or glycols of varying length, all with internal ethylene oxide repeat units.  Although
such slow-forming products should pose no problem in rapid (15-30
second) RIM processes with styrene, utilization in longer diblock or
triblock anionic solution processes may be hindered.  Olefin-termina-
ted degradation products could extinguish growing chain ends via Michael addition, and hydroxyl functional groups by proton transfer.
In addition, linear ethers are more susceptible to n-BuLi degradation
than the cyclic crown ethers, leading to a potential cascading effect
once linear fragments are formed from the macrocycle.[18,37]

The addition of crown ether to the n-BuLi-styrene-benzene system
resulted in dramatic modifications of the polymerization characteristics, supporting the conclusions from the [13]C-NMR stability studies.
Direct comparative studies showed that 15-crown-5 is a more effective
accelerant and $M_v$ builder than 12-crown-4, indicating that contrary to
theoretical calculations, the larger macrocycle more effectively complexes the lithium cation.[6]

Styrene polymerization systems employing 15-crown-5 reached 100%
conversion to polymer in less than 10 seconds, while similar 12-crown-

4 systems required 130 seconds for completion.  The rapid rates confirm the potential of crown ether addition as a route to an anionic
RIM process.  In RIM, two dormant components are usually rapidly mixed
and then injected into a mold.  Reaction time should not exceed one
minute, with < 30 seconds preferred.  In the crown ether accelerated
system, approximately 75% of the reaction mixture is projected to be
filler due to the high heat of reaction and potential shrinkage of
the forming polystyrene.  The filler replaces the benzene used in
the preliminary kinetic studies, and the preferred nature of the material is unsaturated resin prepolymer so that a system concentration
of 2 M in vinyl groups can be maintained.  Due to the rapid kinetics,
the final mixing must be done rapidly.

A conceptual design to accommodate the crown ether accelerant
involves two stage mixing:

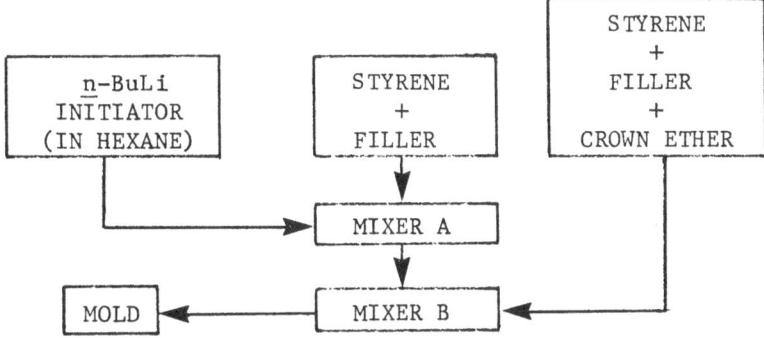

An excess of n-BuLi must be used in order to purge a significant portion of the impurities.  The initiator can be stored in small volumes
of hexane solution since it is unnecessary to have approximately equal
volumes of the two components in Mixer A.  The mixing in A requires
relatively slow speeds since the overall rate is slow without crown
ether present (Table V and Figure 4).

The two inputs to Mixer B should be formulated to be approximately equal in volume and viscosity to facilitate the use of rapid
impingement mixing.  A tubular reactor is currently being designed
and built at Georgia Tech incorporating the two-stage concept.  Additional kinetic studies will be conducted, and benzene will gradually be replaced by resin prepolymer filler to approximate a true
RIM process.

The molecular weight of polystyrene produced from the modified
system increased to an optimum value as the crown ether:n-BuLi molar
ratio approached 0.5:1.  Increasing the ratio toward 1:1 caused a
near symmetrical decrease in the $M_v$.  As the population of crown
ether-complexed chain ends increased relative to noncomplexed ends,

the rapid overall rate of polymerization dictated that the chain
length was limited by monomer supply.

The characteristics of the n̲-BuLi-isoprene-heptane system were
also modified by addition of crown ethers.  Addition of 15-crown-5
reduced the time to complete conversion by 83%, while 12-crown-4 addition dictated a 67% reduction.  Although crown ether:n̲-BuLi ratios
approaching 0.5:1 were prohibited by solubility limitations of the
crown ether complexes in nonpolar heptane, moderate increases in $M_v$
were obtained with both 15-crown-5 and 12-crown-4 by using ratios in
the 0.17-0.32:1 range.  Use of the crown ethers in styrene-diene-sty-
rene or the analogous diblock, commercial anionic systems should substantially improve productivity by reducing the time of reaction for
the slow diene polymerization step.  Manipulation of the crown ether:
metal alkyl ratio should also give interesting property modifications
in the final materials by broadening the polydispersity indices of
the individual styrene and diene blocks.

Polymerizations of Union Carbide's CP grade of butadiene with
the n-BuLi-crown ether complex as initiator were unsuccessful due to
an inability to fully free the monomer from impurities.  Such a system
will require a monomer source with purity standards as rigorous as
those now employed in commercial anionic polymerization processes.[27]
Based on the polyisoprene results, the more reactive butadiene should
present similar or enhanced opportunity for crown ether acceleration
and $M_v$ enhancement provided the monomer purity is vigorously maintained.

No studies were conducted on the effect of crown ether addition on
stereoelectivity or stereospecificity of the styrene and diene polymerization mechanisms.  Halasa and Cheng have reported a higher percentage of 1,2-polybutadiene produced by addition of the high molecular weight tricyclohexyl- or dicyclohexyl- 18-crown-6 ligands to the
n-BuNa initiator system.[19]  No effects on cis-, trans- preference or
on polystyrene tacticity were given.  With the less-bulky, unsubstituted 15-crown-5 and 12-crown-4 ligands and their high binding ability, however, less effect on the mode of monomer addition to the
growing chain end is predicted.  The anion and cation should be highly
solvent-separated, with the cation shielded from the active carbanion
site by the crown ether.  Formation of a "harder" carbanion on complexation should also have little effect on the stereospecificity
or stereoselectivity of the monomer addition.  Further research is
required, however, to conclusively determine the presence or absence
of crown ether effects on the configuration of the polystyrene and
polydiene backbones.

The crown ether system also holds potential in polymerization
of more random SBR rubber.  In batch copolymerization with a combined comonomer feed, the initial portion of the chain is mainly

butadiene. As the diene is exhausted, a block of mainly polystyrene is formed at the end of the chain. Halasa and Cheng copolymerized styrene and butadiene with n-BuNa initiator complexed with substituted crown ethers.[19] Copolymers of high $M_v$, high conversion and no block polystyrene were synthesized. Similar advantages in SBR production are predicted for the n-BuLi/unsubstituted crown ether catalyst systems developed in the current research.

Polymerizations of styrene utilizing "naked" acetate or cyanide/18-crown-6 complexes as initiators instead of n-BuLi gave high $M_v$ products, but low percent conversions. The results were attributed to the low crown ether:salt molar ratio employed (0.07:1) and scavenging of the bulk of the "naked" anion by impurities. The extremely high $\eta_{inh}$ values obtained (2.43-2.46 dL/gm), however, indicated that the anions were sufficiently nucleophilic to initiate the polymerization, and that the salts had potential as replacements for n-BuLi. Further research is required at higher anion concentration and crown:salt molar ratios to optimize the polymer yields and assess the kinetics.

Attempts to modify the bulk anionic polymerization of caprolactam via crown ether addition proved unsuccessful. The 18-crown-6 was apparently undergoing combined thermal and chemical degradation to yield colored fragments. No polymer was isolated from the reactor, indicating that the anion of caprolactam (the polymerization initiator) or of the amide chain was involved in the degradation reactions. Nylon 6 was obtained in high yield and molecular weight from identical systems without crown ether. The results contradict the claim byCollman that the NaOH or KOH/18-crown-6 catalyst system can be used to bulk polymerize caprolactam at high temperatures.[25b]

## Acknowledgement

The support of the National Science Foundation Grant No. ENG-76-10141 for part of this research is gratefully acknowledged.

## References

1. M. Starks and C. L. Liotta, "Phase Transfer Catalysis," Academic Press, New York, NY, 1978.
2. R. M. Izatt et al, Science, 164, 443 (1969).
3. R. G. Ackman, W. H. Brown and G. F. Wright, J. Org. Chem., 20, 1147 (1955).
4. C. J. Pedersen, J. Am. Chem. Soc., 89, 2495 (1967).
5. C. J. Pedersen and H. K. Frensdorff, Angew. Chem., Int. Ed. Eng., 11(11), 16 (1972).
6. H. K. Frensdorff, J. Am. Chem. Soc., 93, 600 (1971).

7.  R. N. Green, Tetrahedron Lett., (18), 1793 (1972).
8.  C. J. Pedersen, J. Am. Chem. Soc., 92, 386 (1970).
9.  C. J. Pedersen, ibid., 92, 391 (1970).
10. C. J. Pedersen, Op. Cit., 89, 7017 (1967).
11. G. W. Gokel, D. J. Cram, C. L. Liotta, F. L. Cook and H. P. Harris, J. Org. Chem., 39, 2445 (1974).
12. G. Johns, C. J. Ransom and C. B. Reese, Synthesis, 515 (1976).
13. F. L. Cook, C. L. Liotta, et al, Tetrahedron Lett., (46), 4029 (1974).
14. F. L. Cook, C. L. Liotta and C. W. Bowers, J. Org. Chem., 39, 3416 (1974).
15. S. Kopolow, T. E. Hogen-Esch and J. Smid, Macromolecules, 6(1), 133 (1973).
16. K. Molinari, F. Montanari and P. Tundo, J. Chem. Soc., Chem. Comm., 639 (1977).
17. M. Tomoi, et al, Tetrahedron Lett., (33), 3031 (1979).
18. A. F. Halasa and T. C. Cheng, U. S. Pat. 3,856,768, Dec. 24, 1974.
19. A. F. Halasa and T. C. Cheng, J. Polym. Sci., Polym. Chem. Ed., 14, 583 (1976).
20. S. Alev, F. Schue and B. Kaempf, J. Polym. Sci., Polym. Lett. Ed., 13, 397 (1975).
21. S. Slomkowski and S. Penczek, Macromolecules, 9(2), 367 (1976).
22. B. Yamada, et al, J. Polym. Sci., Polym. Lett. Ed., 14, 277 (1976).
23. T. Suzuki, et al, ibid., 14, 675 (1976).
24. J. A. Orvik, J. Am. Chem. Soc., 98, 3322 (1976).
25. (a) W. Danowski, W. T. Flowers and R. N. Haszeldine, J. Fluorine Chem., 9, 94 (1977); (b) J. P. Collman, U. S. Pat. 4,073,788, 1978.
26. S. Boileau, et al, J. Polym. Sci., Polym. Lett. Ed., 12, 203 (1974).
27. G. Odian, "Principles of Polymerization," McGraw-Hill, New York, NY, 1970.
28. M. Szwarc, in A. Ledwith and A.M. North, Eds., "Molecular Behavior and Development of Polymeric Materials," Wiley and Sons, New York, NY, 1974, pp. 1-50.
29. D. Worsfold and S. Bywater, Can. J. Chem., 38, 1895 (1960).
30. K. Kircher, S. Schaper, and G. Menges, SPE Tech. Papers, 21, 481 (1975).
31. H. K. Reimschuerosel, in K. Frisch and S. L. Reegen, Eds., "Ring-Opening Polymerizations," Vol. 2, Marcel-Dekker, New York, NY, 1969, p. 303.
32. F. J. Welch, J. Am. Chem. Soc., 81, 1345 (1959).
33. K. F. Driscoll and A. V. Tobolsky, J. Polym. Sci., 35, 259 (1959).
34. D. Brown, et al, "Synthesis of Macromolecular Compounds," Wiley and Sons, New York, NY, 1971, p. 113.
35. H. Gilman and F. K. Cartledge, J. Organomet. Chem., 2, 447 (1964).

36. A. Dabdoub, Ph.D. Thesis with C. L. Liotta, School of Chemistry, Georgia Institute of Technology, Atlanta, GA, 1976.

37. C. E. Coates and W. Wade, "Organometallic Compounds: Vol. 1. The Main Group Elements," 3rd ed., Methuen and Co., Ltd., London, England, 1967.

38. M. Szwarc, "Living Polymers and Electron Transfer Processes," Wiley Interscience, New York, NY, 1969, p. 260.

39. D. H. Richards, in R. N. Howard, Ed., "Developments in Polymerization - I," Applied Science Pub., Ltd., London, England, 1979, pp. 15-16.

40. R. V. Figini, G. Lohr and G. V. Schulz, J. Polym. Sci., B, 3, 985 (1965).

41. L. L. Bohm, et al, Fortschr. Hochpolym. Forsch., 9, 1 (1972).

42. Ref. 38, p. 501.

43. H. Hsieh, C. Vranik and R. Fauar, Phillips Petroleum Research Center, Bartellsville, OK, Private Communication to R. M. Burton, 1978.

44. C. L. Liotta, H. P. Harris, et al, Tetrahedron Lett., (28), 2417 (1974).

45. F. L. Cook, C. L. Liotta, et al, "Organic Synthesis," Vol. 57, John Wiley and Sons, New York, NY, 1978.

# RECENT CHEMICAL AND REINFORCEMENT DEVELOPMENT IN THE REINFORCED

# REACTION INJECTION MOLDING PROCESS FOR AUTOMOBILE APPLICATIONS

P. Z. Han, W. L. Lennon and R. B. Ratajczak

Production Engineering
Oldsmobile Division
General Motors Corporation
Lansing, MI 48921

## INTRODUCTION

Since the early 70's, fascias and bumper covers produced by the
RIM (Reaction Injection Molding) process have been extensively used
in the automobile industry. The advantages of high impact, class
"A" surface and light weight merit consideration for volume usage
by the industry. Yet the application has its limitations, the fascia materials are relatively poor in heat resistance and thermal expansion characteristics. Recent development work to improve the heat
resistance and thermal expansion properties by adding reinforcement
agents in the materials were carried out. At the same time RIM
equipment that can handle abrasive fillers was being developed.
Oldsmobile Division, General Motors Corporation, is the first automobile division to utilize the concept of RRIM (Reinforced Reaction
Injection Molding) in designing a fender application for their 1981
"Sport Omega." This paper is the summary of the development work
in material evaluation for the RRIM fender application. Also, Oldsmobile Division recognizes the limitations of the fascia material
in heat resistance and thermal expansion properties; a RRIM fascia
material development was also carried out within the Division to
improve the quality of the product in fascia application.

## EXPERIMENTAL PROCEDURE AND EQUIPMENT:

The experiment in evaluating RRIM material involved three main
steps: material blending, molding of the part and physical property
testing of the molded part. All experiments were carried out in the
RRIM development area of Oldsmobile Division. First, all materials

evaluated in this experiment were urethane materials. The materials were preblended into two components, there were polyol components containing soft block polyol, hard block extender, catalysts and glass reinforcement, and isocyanate component containing isocyanate alone. The blended components were transferred to the machine holding tanks by the air powered diaphram pumps. The materials in the machine holding tanks were recirculated and temperature conditioned from the tank to the mixing head by the progressive cavity pumps. The materials were constantly recirculated while the machine was idling to prevent the settling of the glass reinforcement. Second, part molding involved establishing the optimum molding conditions such as reaction ratio, injection rate, injection pressures, material temperature, mold temperature and mold residence time, molding and inspecting of the part, and handling of the molded part before testing. A typical molding cycle consists of applying the mold release, closing of the mold, injecting the materials, curing of the material in the mold, opening the mold and removing the part. A mold release had to be applied during each of the molding cycles because the urethane chemical used in the RRIM process is a good bonding agent. All molded parts were placed in a convection oven at 250°F for one hour to further crosslink the unreacted isocyanate. Third, all test methods used in this equipment were standard ASTM methods. All test samples were preconditioned in a 75°F, 75% RH room for 24 hours before testing. Test samples parallel to the material flow and perpendicular to the material flow were obtained from predetermined locations. The reinforcement agent such as milled glass fiber has an aspect ratio between 9 to 26, depending on the fiber length, and tends to orientate itself when injected into the mold. The parallel and perpendicular results may vary depending on the distance that the injected material may flow in the mold.

RESULTS AND DISCUSSION

Chemical Reactivity

Chemical reactivity of a material was an important factor in selecting a production material for the production of the RRIM "Sport Omega" fenders. The chemical reactivity affects the rate that production parts can be produced; the higher the chemical reactivity, the higher rate production parts can be produced. Therefore, during the evaluation, chemical reactivity of a material was determined. The method for determining the chemical reactivity was to judge the ability of the material to fill the mold at a common injection rate. The fast reactive material tends to retard the flowability due to the gel viscosity buildup. In the reactive study, five different material systems with various components that will affect the reactivity were evaluated. Table I shows the components breakdown of these five material systems. After the evaluation, it was found that

TABLE I

Chemical Components in Chemical Reactivity Study

| Material Formulation Number | Soft Black Polyol | Hard Block Extender | Catalyst | Isocyanate |
|---|---|---|---|---|
| 101 | Polyether | Diamine | Tin | MDI Prepolymer |
| 119 | Polyether | Diamine/Ethylene Glycol | Tin | Polymeric MDI |
| 107 | Polyether | Ethylene Glycol | Tin/Amine | Polymeric MDI |
| 104 | Copolymer | Ethylene Glycol | Tin | Polymeric MDI |
| 117 | Copolymer | Ethylene Glycol | Tin | Liquefied MDI |

the material containing diamine extender was the fastest in reactivity. The material containing tin and amine synergistic catalysts
was slower and material containing tin catalyst alone was the slow-
est. Also, during the evaluation of chemical reactivity of isocyanates, it was found that alkyl phosphate liquified MDI was more reactive than prepolymeric MDI and more reactive than polymeric MDI. It
can be concluded that the material with the fastest reactivity is
made of a diamine hard block extender, a tin and amine catalyst synergistic mixture and an alkyl phosphate liquified MDI. But in selecting a production material, the ability to produce more parts per
day was not the only factor to be considered; the ability to produce
more parts with good surface quality for finishing and the ability
to meet the on-car test performance requirements were also judged.

## Effect of Glass Reinforcement on Physical Properties

Physical properties were obtained from the parts molded during
the chemical reactivity study. The results were used to determine
the effect of glass reinforcement on physical properties. Table
II shows the physical property comparison of the glass reinforced
materials at 20% level with their respective material unreinforced.
As expected, the glass reinforcement does affect all the physical
properties of the material, their densities were increased, their
flexural moduli were increased, their heat sag and linear coefficient of thermal expansion were improved and impact was reduced.
The improved heat resistance and linear coefficient of thermal expansion indicate that the "Sport Omega" fender production material
has to be reinforced. Also, as was expected, the physical properties
from the test samples perpendicular to the flow of the material and
test samples perpendicular to the flow of the material showed considerable difference; the parallel values were much higher than the
perpendicular ones. In the fender application, the difference did
not affect the on-car performance of the material. Also, this difference can be minimized by reducing material flow distance in the
mold, or by reducing the injection rate, or by reducing the chemical
reactivity, or by using a different reinforcement agent.

## TABLE II

### Effect of 20% Glass Fiber Reinforcement on Physical Properties

| | OLDS-RIM-101 // L | OLDS-RIM-119 // L | OLDS-RIM-107 // L | OLDS-RIM-104 // L | OLDS-RIM-117 // L |
|---|---|---|---|---|---|
| Density g/cc | 1.12-1.19 (1.06) | 1.07-1.11 (1.00) | 1.11-1.19 (1.00) | 1.13-1.20 (1.01) | 1.19-1.29 (0.98) |
| Tensile Strength psi | 3300  3400 (4000) | 3800  4200 (4130) | 4900  4800 (4500) | 5300  4400 (4500) | 5500  5000 (4300) |
| Tensile Elongation | 220  240 (280) | 30  30 (130) | 16  26 (90) | 20  40 (89) | 25  30 (123) |
| Flexural Modulus Mpsi RT | 191  82 (53) | 241  108 (93) | 306  179 (120) | 331  152 (115) | 300  170 (92) |
| -20°F | 298  189 (119) | 396  187 (179) | 430  266 (245) | 512  261 (235) | 490  270 (179) |
| 158°F | 132  54 (35) | 167  70 (60) | 197  99 (70) | 218  99 (63) | 220  100 (59) |
| Heat Sag Inch | 0.1  0.4 | 0.1  0.2 | 0.3  0.5 | 0.1  0.3 | 0.0  0.2 |
| LCTE | 36  69 (97) | 30  62 (95) | 32  55 (76) | 15  37 (70) | 28  42 (65) |
| Notched Izod Ft lb/in | 5.3 (9.6) | 3.5 (5.2) | 2.5 (4.0) | 4.1 (4.7) | 5.2 (6.5) |

(    ) Physical property of the corresponding unreinforced materials.

## Effect of Glass Reinforcement Percent on Physical Properties

Since it was determined that glass reinforcement improved the heat resistance and thermal expansion properties and it was known that fascia materials are notoriously poor in heat resistance and thermal expansion properties, development work has been carried out to improve the heat resistance and thermal expansion properties by adding glass reinforcement in the fascia material. Using a typical fascia material of 25,000 psi flexural modulus, 2%, 5%, 10%, and 20% levels of glass reinforcement were added. Parts were molded and physical properties were determined. Due to part design, no perpendicular property data was obtained. Figure 1 shows that the glass reinforcement reduced tensile elongation from 330% at unreinforced to 185% at 20% loading. Figure 2 shows that the glass reinforcement improved the heat resistance property from 0.23 inches heat sag of

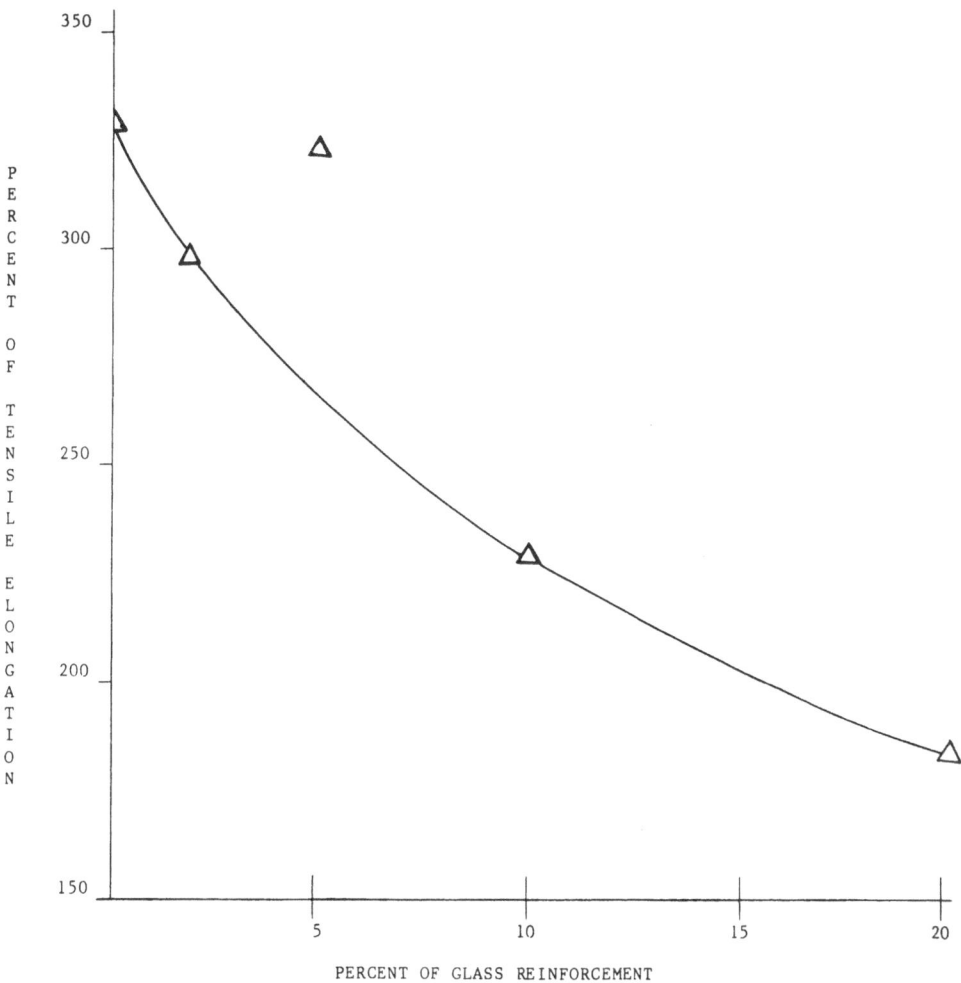

Figure 1. Effect of glass reinforcement on percent of elongation.

the unreinforced to 0.07 inches heat sag of the 20% loading.  Figure 3 shows that the glass reinforcement improved the linear coefficient of thermal expansion from 95 x 10 inch/inch/°F of the unreinforced to 35 x 10 inch/inch/°F of the 20% loading.  Finally Figure 4 shows that the glass reinforcement increases flexural modulus from 24,000 psi of the unreinforced to 78,000 psi of the 20% loading.

In the fascia application, flexural modulus and impact proper- ties are critical.  The fascia material has to pass the 5 mph impact requirement and still cause no damage to the adjacent body panels. Therefore, it is important to develop a reinforced fascia material

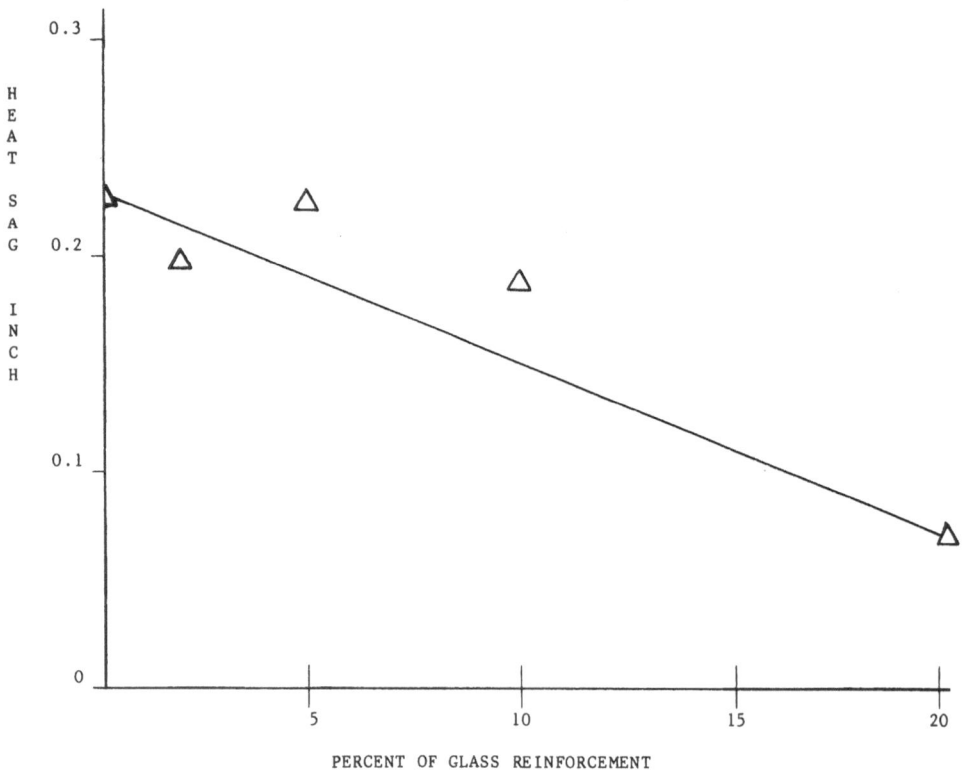

Figure 2. Effect of glass reinforcement on heat sag.

possessing improved heat resistance and thermal expansion properties
without affecting the flexural modulus and impact properties. One of
the solutions was to reinforce a fascia material at lower flexural
modulus levels and to use the glass reinforcement to obtain the desired flexural modulus. The following example shows that the heat
resistance and thermal expansion properties can be improved without
affecting the flexural modulus property:

| Reinforcement % | 0 | 10 |
|---|---|---|
| Flexural Modulus MPSI | 25 | 25 |
| Tensile Elongation % | 270 | 240 |
| Heat Sag - inch | 0.24 | 0.12 |
| LCTE x $10^{-6}$/inch/inch/°F | 96 | 50 |

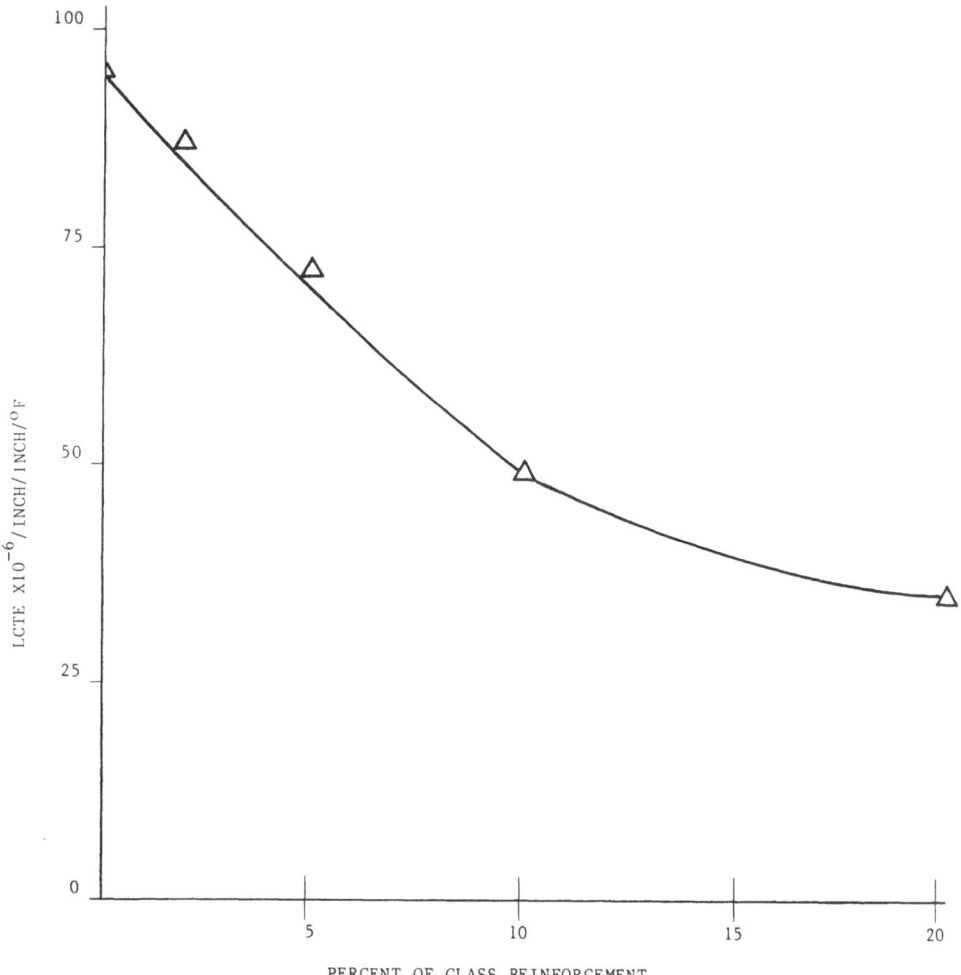

Figure 3.   Effect of glass reinforcement on linear
coefficient of thermal expansion.

The adding of a glass reinforcement to fascia materials to improve
their heat resistance and thermal expansion properties certainly
merit further investigation.

## Effect of Types of Glass Reinforcement on Physical Properties

In the RRIM process, it is important to find a glass reinforcement that will produce maximum physical properties in a RIM material
without causing processing problems.  Six different milled glass
fibers and two chopped glass fibers were investigated.  Tables III

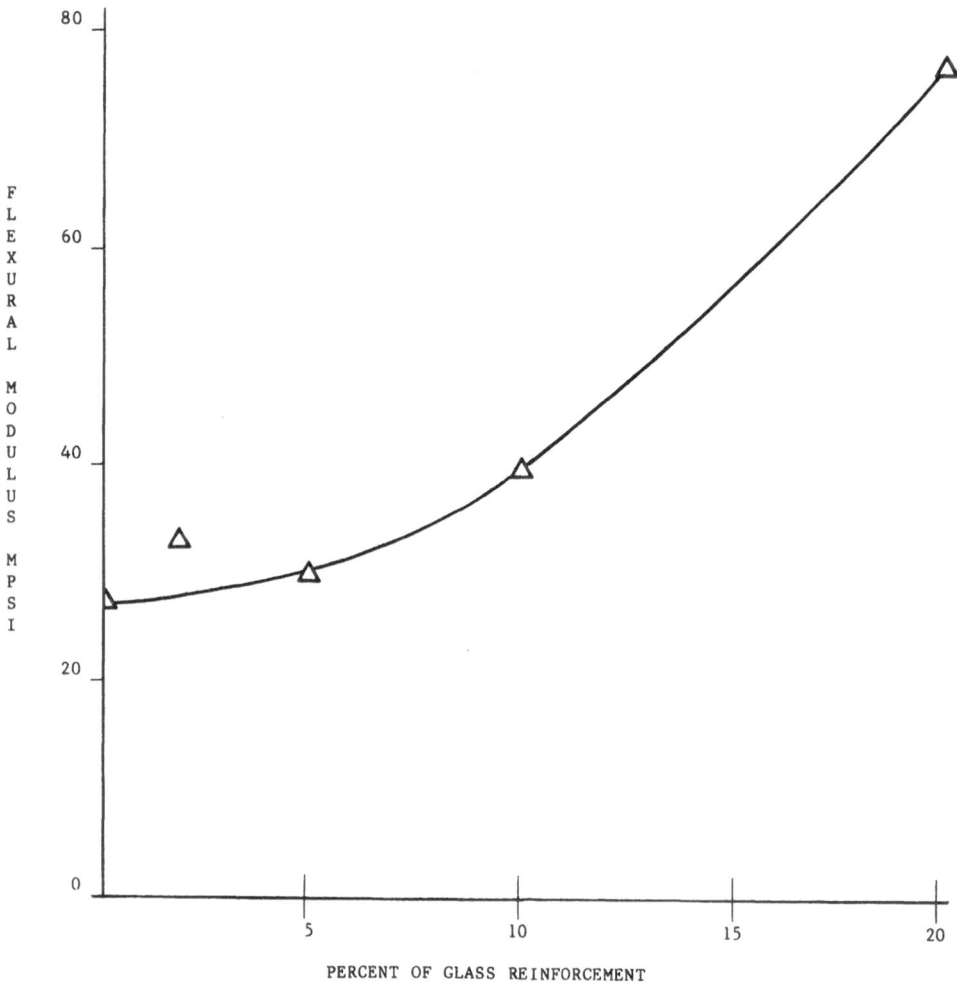

Figure 4.   Effect of glass reinforcement on room
            temperature flexural modulus.

and IV show the comparison of their LCTE and flexural modulus properties data.  Results indicate chopped glass fibers produced better
physical properties than milled glass fibers, but chopped glass
fibers were very difficult to process due to the viscosity buildup.
Results also indicate the milled glass properties are consistent
with their responding fiber fize; longer milled glass fibers yield
higher physical properties.  In the process evaluation, all milled
glass fibers were processed without any difficulty, except the 1/8"
milled glass fiber II which acted like the chopped glass fiber.

TABLE III

## Effect of Glass Fiber Size on Flexural Modulus

| Glass Fiber Reinforcement Screen Size or Chopped Length | | Flexural Modulus $10^3$ psi | | | | | |
|---|---|---|---|---|---|---|---|
| | | Parellel | | | Perpendicular | | |
| | | RT | -20 | 158 | RT | -20 | 158 |
| Milled Glass I | | | | | | | |
| 1/32" | (17.5%) | 230 | 380 | 120 | 150 | 280 | 60 |
| 1/16" | (19.6%) | 260 | 450 | 160 | 150 | 300 | 90 |
| 1/8" | (21.1%) | 320 | 520 | 180 | 170 | 310 | 80 |
| Milled Glass II | | | | | | | |
| 1/32" | (20.3%) | 350 | 540 | 220 | 160 | 300 | 90 |
| 1/16" | (20.4%) | 350 | 550 | 220 | 170 | 300 | 90 |
| 1/8" | (8.3%)* | 230 | 390 | 140 | 150 | 270 | 90 |
| Blend I and II | | | | | | | |
| 1/16" | (19.9%) | 340 | 540 | 220 | 200 | 330 | 100 |
| Chopped Glass I | | | | | | | |
| 1/16" | (7.7%)* | 220 | 340 | 140 | 130 | 220 | 70 |
| Chopped Glass II | | | | | | | |
| 1/16" | (6.5%)* | 220 | 310 | 150 | 130 | 220 | 70 |

*Viscosity too high for processing.

TABLE IV

## Effect of Glass Fiber Size on Linear Coefficient of Thermal Expansion

| Glass Fiber Reinforcement Screen Size or Chopped Length | | Coefficient of Thermal Expansion of | | | |
|---|---|---|---|---|---|
| | | Parallel | | Perpendicular | |
| | | -20-70 | 72-175 | -20-70 | 72-175 |
| Milled Glass I | | | | | |
| 1/32" | (17.5%) | 29.6 | 26.2 | 48.7 | 53.3 |
| 1/16" | (19.6%) | 29.0 | 27.9 | 47.1 | 57.1 |
| 1/8" | (21.1%) | 24.3 | 24.4 | 48.7 | 56.7 |
| Milled Glass II | | | | | |
| 1/32" | (20.3%) | 30.8 | 17.7 | 54.6 | 55.6 |
| 1/16" | (20.4%) | 23.5 | 22.1 | 54.3 | 54.9 |
| 1/8" | (8.3%)* | 35.3 | 29.3 | 51.7 | 62.2 |
| Blend I and II | | | | | |
| 1/16" | (19.9%) | 29.2 | 26.4 | 48.2 | 56.0 |
| Chopped Glass I | | | | | |
| 1/16" | (7.7%)* | 38.1 | 30.2 | 58.3 | 62.5 |
| Chopped Glass II | | | | | |
| 1/16" | (6.5%)* | 28.6 | 27.9 | 51.7 | 61.1 |

*Viscosity too high for processing.

OTHER RRIM MATERIALS

Although all the experiments in this paper were based on the urethane materials, the potential of other materials that can be reaction injection molded are unlimited. Recently, nylon, epoxy and esters RIM materials started to surface. Development work has to be carried out between the chemical companies and equipment companies to make these materials production worthy.

PROCESS IMPROVEMENT FOR RRIM

In the RRIM process, a mold release has to be applied on the mold during each of the molding cycles, the applying of the mold release reducing the production rate. Also, most mold releases are wax-based, using a solvent carrier which has emission implications. Therefore, elimination of external mold release is definitely a process improvement. Also in the RRIM process, all molded parts have to be cured at 250°F for one hour before the physical properties are completely secured. This step is time consuming and energy wasting. Elimination of post-cure steps will be another improvement. Materials with better physical properties, such as heat distortion property, will be a big plus in the RRIM process. The material with better heat distortion will eliminate the expensive support fixtures that are used during post-curing and painting. Finally, in the RRIM process, two components are being used. In order to obtain maximum physical properties, accurate metering and mixing have to be employed. One component RRIM material will be a big advantage because expensive equipment demands are reduced.

CONCLUSIONS

RRIM process has growth potential. Work has to be done in the areas of new material development, new process development and new equipment development to promote expanded use of the RRIM or RIM process.

Acknowledgements

The authors would like to extend their appreciation to the chemical companies who furnish the materials to make this study possible. The authors would also like to extend their appreciation to Owens Corning Fiberglas for their assistance in the glass reinforcement development work.

# FIBER GLASS REINFORCED REACTION INJECTION MOLDING

M. M. Girgis, B. Das and J. A. Harvey

PPG Industries, Inc.
Fiber Glass Research Center
Pittsburgh, PA 15230

## INTRODUCTION

Reaction injection molding is a fast "in situ" polymerization process which forms the desired polymer from its monomeric liquids. The process is characterized by the use of high-speed impingement to mix reactive monomers or prepolymers which rapidly polymerize in a heated mold under low pressure. The flow rate of two such reactive monomers, polyol and diisocyanate, must be measured and controlled to provide the proper stoichiometrical ratio for the resulting polyurethane. A schematic diagram of the RIM process is shown in Figure 1.[1] The liquid reactant streams which are rapidly mixed in the mixhead are fed into a heated mold through a runner and react to form a solid part.

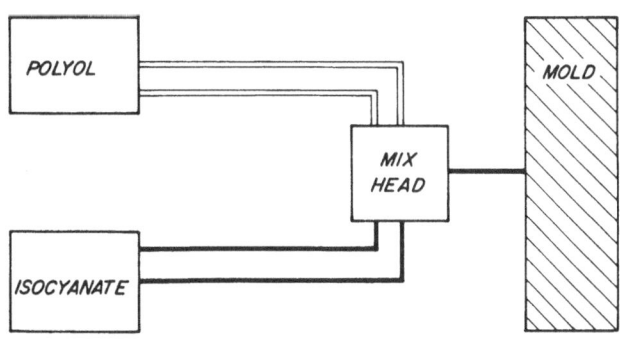

Figure 1. RIM urethane process.

Reaction injection molding, RIM, is a relatively new process which has made significant advances in exterior auto parts and other non-auto markets since RIM urethane bumper covers and flexible fascia made their appearance on American cars in 1975. The incentive for this rapid growth stems from:

1. The federally mandated front and rear end non-damageability standards which suggest the implementation of elastomeric bumper covers, and

2. The low temperature and pressure requirements of producing a RIM part. These requirements give RIM the advantage of having lower energy requirements per unit in comparison to parts made from other materials (Figure 2).

The main disadvantage of RIM parts has been attributed to the soft elastomeric nature of the final polyurethane. Another disadvantage to acceptance of RIM urethane for large auto body parts is due to its high degree of thermal expansion. Currently, resin manufacturers working on modifications of the resin system have provided polyurethanes with flexural moduli in excess of $4 \times 10^5$ psi and with acceptable impact properties; but these modifications in formulation have not yet solved the problem attributed to the large difference in thermal expansion between the polyurethane part and the steel framework to which it is attached. Also, part distortions due to creep and high temperature modulus have been observed in paint bake

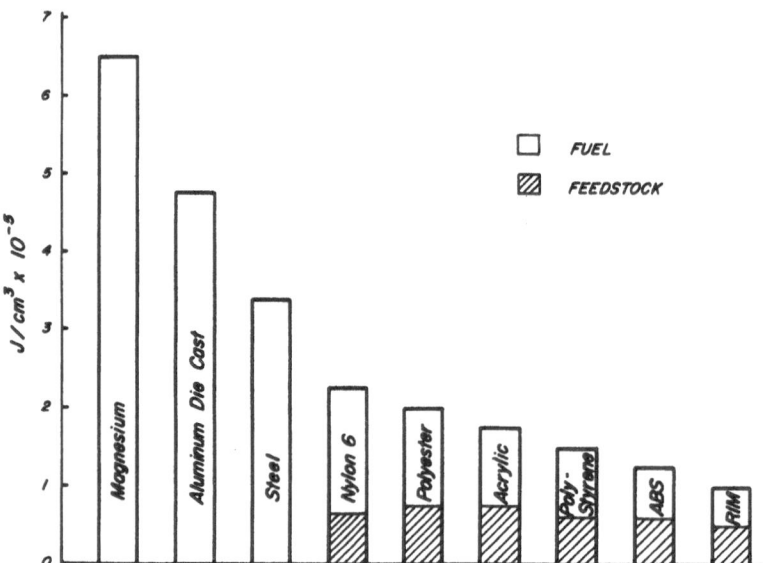

Figure 2. Energy requirements to produce and form various materials.

ovens after 30 minutes.  Thus, reinforcement with temperature insensitive and high modulus materials, such as fiber glass, is considered essential in auto body and other large rigid applications.

Milled fiber glass is currently considered to be the preferred
RIM reinforcement due to its compatibility with appropriate RIM
machines and its lesser degree of degradation of impact properties
when compared to other inorganic fillers.  Reinforced RIM polyurethane using 1/16" milled fiber glass at 25% or more loadings has
been achieved in a high modulus ($1.5 \times 10^5$ psi) urethane formulation.
Such milled fiber loadings produced composites with a flexural modulus of $3 \times 10^5$, an elongation of 20%, an Izod impact of 3 ft lb/in,
a sag of 0.1 inch for 4-inch specimen at 325°F for 30 minutes, and
a thermal expansion coefficient of $20 \times 10^{-6}$ in/in°F, all properties
being measured along the flow direction.  It has been found that
there is a considerable degree of fiber orientation along mold flow
lines, thus leading to an approximately 2:1 enhancement of properties
in that direction.

The use of chopped fiber glass strand bundles as reinforcing
elements in RIM urethane systems has not been fully explored, even
though the aspect ratio (length/diameter) is reasonable good for reinforcement.  Preliminary work has shown that physical properties
of such chopped strand reinforced RIM urethane panels are comparable
or superior to milled fiber panels.  This presentation will deal
with the rheological behavior of chopped fiber glass bundles and
filaments in a RIM urethane system, and the machine processing and
properties of reinforced RIM composites.

The relatively limited data on chopped glass fibers in the reinforced RIM urethane system is due mainly to the processing difficulties in current RIM machines.  The major problems are fiber handling (such as fiber dispersion, mixhead abrasion, and moisture pickup
by glass fibers), and an inability to process longer glass fibers
because of the extremely high viscosity of the slurry.  It has been
found that the viscosity of the slurry increases steeply with fiber
content, particularly under low shear conditions, even for the short
chopped fiber length of 1/8 inch (3 mm).[2]  The observed high viscosities and shear thinning behavior can be attributed to fiber network
formation which varies strongly with fiber length.  Gelling in the
mix tanks and lines can occur at fiber concentrations as low as 5%
to 10%.  Fiber settling and compaction at flow restriction points
(as in needle valves and line heat exchangers) can lead to processing
difficulties at low concentrations as well.  The use of 1.5 mm long
fibers for RRIM at a concentration of 8% by weight has been reported
to give properties equivalent to 25% milled fibers.[3]  However, surface quality and production worthiness have not yet been demonstrated.

EXPERIMENTAL

Rheology Study

The polyol used in this study was Mobay Chemical's Bayflex 110-50 system. Glass fibers in the form of integral cut strand, dispersed filaments, and milled fibers were mixed with the polyol. In the case of dispersed filaments, filamentizable chopped strand was stirred with the polyol until the strands had completely converted to dispersed glass filaments. The glass fibers used were 3 mm integral chopped strand containing 400 filaments (12 μm in diameter), and milled fibers (1.6 mm mesh size).

A capillary rheometer (Burrell Sever Model A-120) was used to obtain the rheological data. This consists of an air pressurized (Filtered dried air) 400 cc cylinder leading into a capillary orifice (radius 0.164 cm and length 5.00 cm). Mass flow rates of the slurries were determined at pressures ranging from 20 to 80 psi. The data were obtained at ambient temperatures ranging from 23°C to 26°C, due partly to frictional heating in the flow of the slurries through the capillary. These temperature variations were considered small enough not to have any significant effects on the viscosity of the slurries.[2]

Machine Processing and RRIM Urethane Composites

The VR-75 RRIM machine made by Accuratio Systems, Inc. (ASI) was used in this study. It has 38 liter mix tanks, an air driven agitator and needle valves. The machine also features short, smooth, relatively straight material lines; high pressure metering cylinders; and a mixhead with wear-hardened critical parts such as orifice bars. The more powerful DC-drive agitator and the hydraulic back pressure regulator options would have been preferable but were not available at the time of this study.

A simplified piping diagram of the VR-75 machine is shown in Figure 3. Two hydraulic pumps feed the homogenized chemical reactants into the metering cylinders which form, together with an hydraulic unit, the metering system. During the filling process, each component is recirculated through the closed mixing head and back into the flow line leading to the tanks. When the metering cylinder is filled, the mixing head control piston opens the orifice bars leading into the mixing chamber, permitting the isocyanate and filled polyol to enter at predetermined ratios for impingement mixing. As soon as the shot time has ended, the mixing control piston advances to close the mixing chamber, pushing the mixture into the mold and actuating the wipe piston to clean the chamber.

High pressure recirculation through the mixing head and orifice bars before and after the mix/pour operation minimizes lead lag

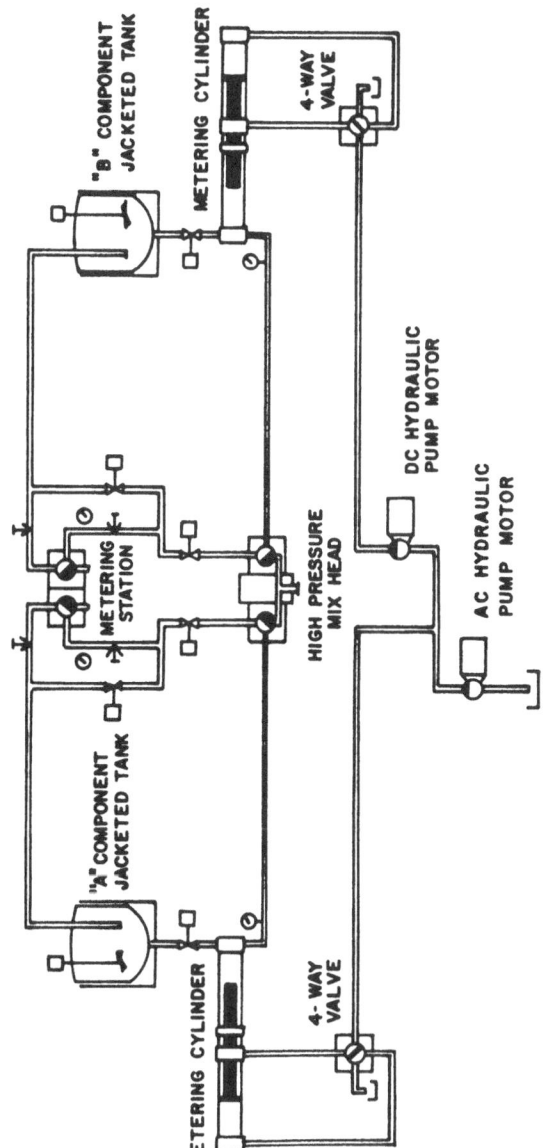

Figure 3.  Simplified Piping Diagram, VR-75 RRIM machine.

problems and ensures safe processing without clogging orifices with fibers or reactive materials.

An aluminum plaque mold (250 x 250 x 3 mm) with an aftermixer section was used. It was equipped with cores for hot water circulation and was controlled at 60°-65°C. The clamping force was supplied by a simple air bag press. The resin used was the Bayflex 110-50 system from Mobay Chemical Corporation, which consists of a polyurea dispersion in a blend of polymeric and short chain diols ("B" side), with a specific gravity of 1.02 at 30°C and a viscosity of 800 MPa's (cp). An MDI type isocyanate having a specific gravity of 1.20 at 30°C and a viscosity of 350 MPa's was used for the "A" side at a volumetric ratio B:A of 2:1 for the unfilled system. The liquids were kept at 32°C to help minimize strand filamentization.

The fiber glass used consisted of a high integrity strand chopped to 3 mm lengths. The glass binder had been designed to provide good resistance to mechanical shearing to produce finely chopped strand. It was not particularly resistant, however, to dissolution by warm polyols and isocyanates. Adequate temperature control of the two liquid reactants is therefore important, particularly in view of the significant heat buildup generated during material recirculation.

Reinforced RIM urethane plaques containing 4.5% and 12.5% by weight of the chopped strand bundles were prepared. The panels which were postcured at 120°C for one hour were found to be translucent and showed clearly the basic, random disposition of the strands in the urethane composite.

RESULTS AND DISCUSSION

In Figure 4, the viscosity vs. fiber concentration data are shown for slurries of 3 mm chopped filaments, 3 mm chopped strand bundles, and 1/16" milled fibers, at shear rates similar to those attained in the recirculation mode of some RIM machines ($10^3 sec^{-1}$).

It is evident from this data that the viscosity of the polyol slurry builds up quickly in the case of 3 mm (1/8") chopped filaments, and it almost doubles by increasing the concentration from 5% to 10%. In the case of chopped strands, however, the rate of viscosity increase with glass concentration is significantly lower than that of chopped filaments. It has been found that a 15% to 20% concentration (on polyol) of the integral strand, or a 5% to 10% concentration of chopped filaments, leads to viscosities which are comparable to 70% to 80% (on polyol) of milled fibers at $10^3 sec^{-1}$ shear rate. At the higher shear rates of injection mode, the longer fibers (or strand bundles) produced a lower mix viscosity because of the shear thinning characteristics of chopped strands or fibers.

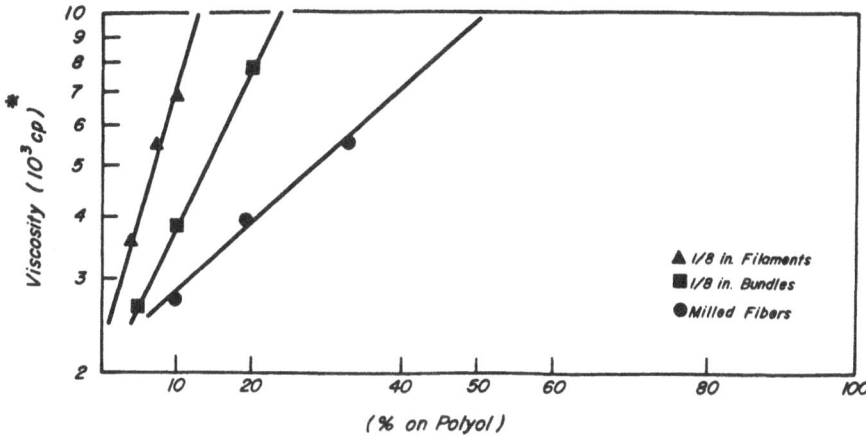

Figure 4.  Viscosity vs. fiber concentration in a RIM polyol.


     The effects of fiber glass type and level on the flexural mod-
ulus of urethane composite are presented in Figure 5.  They indicate
that chopped strand bundles provide an exponential increase in flex
modulus after a "critical" concentration of about 10%, when they
become more efficient in reinforcement than milled fibers.

     Figure 6 shows the superiority of chopped strands in heat sag
over that of milled fiber at a much lower concentration of glass
reinforcement.  At a 12.5% concentration of 3 mm chopped strand
bundles in urethane composite, heat sag (measured by putting 102 mm
cantilever specimens in an oven at 162°C for 30 minutes) is completely eliminated.

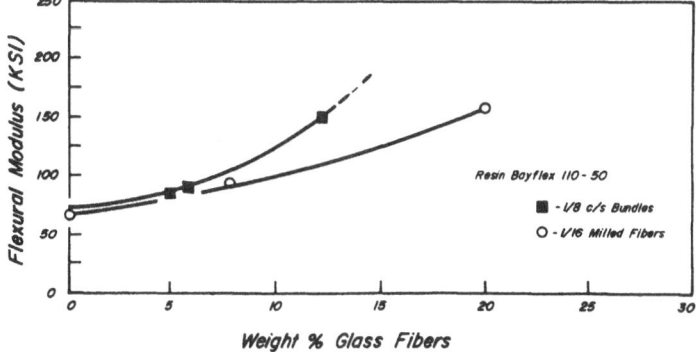

Figure 5.  Flex modulus vs. % glass fibers
           (c/s bundles vs. milled fibers).

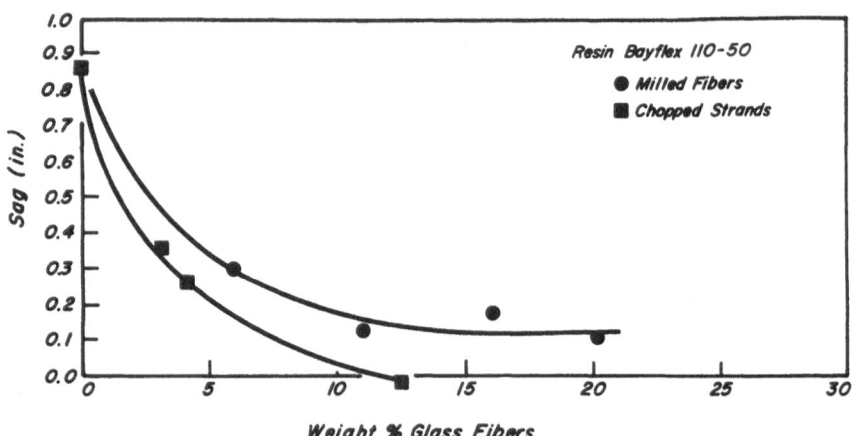

Figure 6.   Heat sag (325°F-1/2 hour) vs. % glass fibers
(c/s bundles vs. milled fibers).

The coefficient of thermal expansion is a measurement of dimensional stability over a wide temperature range (23°C to -40°C). Figure 7 (A & B) shows the effects of fiber glass concentration on the coefficient of thermal expansion measured parallel and perpendicular to the flow direction. It appears from these graphs that 8% to 10% chopped strand is about equivalent to 25% of milled fibers, and there is a degree of anisotropy favoring the chopped strand bundles over milled fibers.

The impact strength of the RRIM urethane composites as measured by Izod indicates, as shown in Figure 8, that chopped strand reinforced panels have significantly higher notched Izod values than those reinforced with milled fibers. However, if impact strength is measured by the falling weight test (Gardner), the situation will be reversed, i.e., 20 vs. 40 in-lb. It is not certain which test is really assessing the true impact strength of the RRIM composite. The Izod test indicates the energy required to break notched specimens under standard conditions. It is calculated as ft-lb per inch of notch and is usually calculated on the basis of a one-inch specimen. The Gardner test, on the other hand, measures the impact energy which produces 50% failures; a failure is when the impact area shows some evidence of composite deformation causing a fine crack to develop on the opposite side of the contact area. It seems that the significance of Izod or Gardner impact values depends mainly on the application, i.e., the impact failures of the RRIM parts in actual service.

Table I summarizes data on chopped strand bundles reinforced panels at the highest loading studied (12.5% by weight) and offers a comparison with available data on panels containing milled fibers

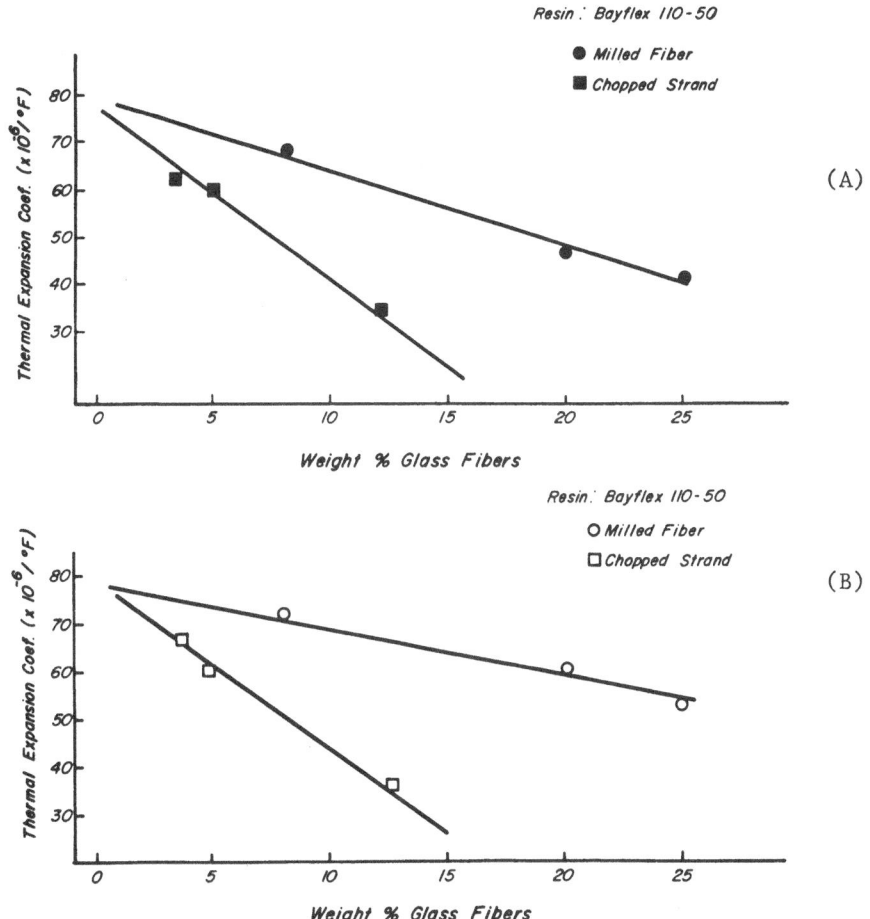

Figure 7.    Thermal expansion coefficients vs. % glass fibers parallel
to flow (c/s bundles vs. milled fibers).

at a 25% level.[4]  The data reflect measurements on specimens with
a parallel to flow orientation.  The target properties listed are
those specified by the automotive industry.[5]  The results suggest
that chopped strand panels are superior in heat sag and Izod impact
values, and comparable in thermal expansion and flexural modulus when
compared to milled fiber panels having twice as much glass reinforcement.


CONCLUSIONS

The combination of processing advantages - lightweight, superior impact resistance; adequate rigidity; and apparently acceptable

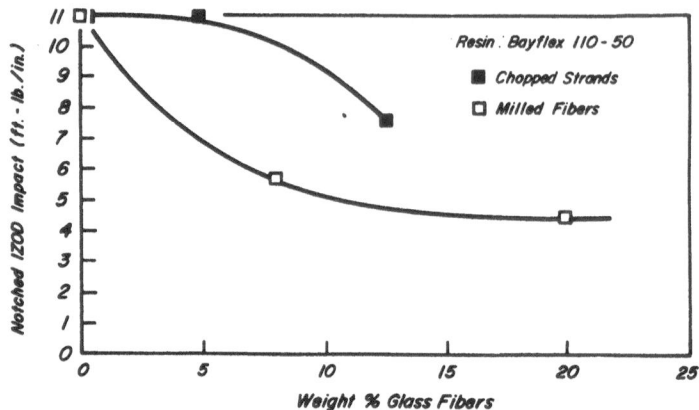

Figure 8.   Impact strength (Izod) vs. % glass fibers (c/s bundles
            vs. milled fibers).


thermal properties make RRIM an attractive candidate for auto body
panels such as fenders.

    At one-half the loading, 1/8 inch (3 mm) chopped strand RRIM
can be competitive in physical properties with 1/16 inch (1.6 mm)
milled fiber RRIM and is superior in some ways (heat sag, Izod
impact).  Impact at low temperature should be obtained.  RRIM using
1/16 inch (1.6 mm) chopped strand should be explored further and
compared with the foregoing.

    A binder with greater solvent resistance must be developed for
the chopped strand bundle product to make it production worthy.


Table I.   Panel Properties.   Chopped Strand Bundles vs. Milled Fibers.

|  | RESIN[1] | 12.5% c/s | 25% MF | TARGET |
|---|---|---|---|---|
| FLEXURAL MODULUS MPa (Ksi) (ASTM D 790) | 340 (50) | 1000 (150) | 1100 (160) | 1000 (150) |
| IMPACT STRENGTH IZOD ( NOTCHED) J/m (ft.-lb./in.) (ASTM D 256) | 590 (11) | 400 (7.5) | 210 (4) | 130 (2.5) |
| THERMAL EXPANSION COEFFICIENT (CLTE) ppm/°C (°F) (ASTM D 696) | 160 (90) | 58 (33)* | 73 (41) | ≤ 50 (30) |
| HEAT SAG (162°C - 30 min.) mm (in.) (ASTM D 3769) | 15 (0.6) | 0 | 2.5 (0.1) | <12 (0.5) |

[1] BAYFLEX 110- 50

* CLTE: 23°C TO -40°C

The processability and surface quality of chopped strand RRIM in short lengths, in the form of bundles, and as dispersed filaments must be carefully evaluated. Any reduction in rating from Class A must be counteracted through appropriate filler use or resin chemistry modification, if a chopped strand product totally competitive with milled fibers is to be achieved.

## Acknowledgments

We would like to express our appreciation to J. R. James, President, ASI, for the use of his facilities to conduct our trials and to W. D. Templin, Sr., Engineer, Mobay Chemical Corporation, for his assistance in physical property determinations, and to I. A. Zlochower for obtaining the experimental data used in this study. We also wish to thank J. Maaghul, Director, Research, Fiber Glass Division, PPG Industries, Inc., for permission to publish this paper.

## References

1. Young, J., Autoproducts, June, 1975.
2. Zlochower, I. A., Das, B., and Hudson, H. J., "Fiber Glass/RIM Urethanes: Rheology vs. Reinforcement," SPI Reinforced Plastics/ Composites Institute, 34th Annual Conference, 1979 (11-C).
3. Chiswall, B. C., and Thorpe, D., "RRIM - A Novel Approach Using Fibreglass," SPI Reinforced Plastics/Composites Institute, 35th Annual Conference, 1980 (22-A).
4. Prepelka, D. J., "Newest Trends in RIM and RRIM - Raw Materials," New Advances in RIM Seminar, Technomic Publishing Company, Pittsburgh, September, 1979.
5. Nelson, G. V., and Dabrowski, A. J., "Designing with RRIM," SAE Technical Paper Series, No. 800513, Detroit, February, 1980.

# REFINING OF THE RRIM PROCESS, MATERIALS

# AND EQUIPMENT IN THE AUTOMOTIVE INDUSTRY

Michelle J. Mikulec

Plastics Development Application Office
Ford Motor Company
Detroit, Michigan

## INTRODUCTION

United States automobile manufacturers are faced with the problem of providing a sufficient proportion of five- and six-passenger cars to satisfy the needs of American families while meeting the requirement for maximum fuel economy dictated by Federal regulations and the rapidly escalating cost of gasoline.

Part of the solution lies in new technologies, such as Ford Motor Company's automatic overdrive transmission and electronic engine controls, along with downsizing to provide cars that are smaller on the outside, but which retain usable passenger space. However, the total solution requires the substitution of lightweight materials, which can hold to a minimum the need to downsize.

The major lightweight materials involved in weight reduction will be: high strength steel, aluminum, thin glass and plastics. Exterior body panels are a natural target for weight reduction because they represent a significant portion of the total vehicle weight.

Recent efforts have been concentrated on development of RRIM technology to be used for exterior body panels. The materials need to have these key physical properties: (1) automotive Class "A" surface finish, (2) acceptable heat stability in car position while being processed through assembly plant ovens up to four times at 285°F - 300°F surface temperatures, (3) the material's heat stability combined with good impact properties at 72°F and -20°F, (4) material stability that assumes no permanent cumulative expansion or contraction will occur when other production steps are taken, (5) capability

231

to be painted with standard body enamel paint in car position.

I want to stress that this paper exclusively discusses the requirements for urethane or non-urethane RRIM materials for the on-line painting process (higher paint curing temperature) rather than the off-line painting process. The materials for off-line painting process (lower paint curing temperature) are here and are being used by the industry.

The requirements for the on-line painting are more difficult to obtain with polyurethane, therefore, non-urethanes are being investigated.

This presentation is centered around REFINEMENTS of the RRIM material, process and equipment development. It is based on my paper "RRIM - A New Process in the Automotive Industry." The rate of RRIM development becomes obvious with a comparison of these two papers.


RIM + RRIM - THEIR POSITION IN THE AUTOMOTIVE INDUSTRY

The RIM process is not new to the automotive industry. Mustang and Capri carlines have front and rear fascias produced with low modulus RIM polyurethane. Off-line part painting is employed.

A new challenge is to produce large exterior parts painted with the car bodies on assembly lines to eliminate the investment cost for off-line painting systems and to insure good color match between the plastic and steel body parts. This challenge leads to the new technique called RRIM. The RRIM process consists of adding a filler to the polymer to increase its heat stability, stiffness and coefficient of linear thermal expansion (CTE) in comparison to the RIM process. The addition of the abrasive filler to material tanks brings about a need for process, equipment and filler development. Each of these will be discussed individually.


RIM + RRIM - ENERGY CONSERVING PROCESSES

Conservation of energy is becoming more and more important not only for consumers but also for the automotive industry in general.

The RRIM process is competing with the metal stamping and SMC processes in the area of exterior body panels. One must consider the energy aspect of a new process and compare it with well-established processes. It is expected that energy consumption will play a significant role when new production facilities are planned. The needed energy can be divided into two categories: energy of the feed stock the part is made from, and energy from the fuel required to process the part. Metal materials do not have an energy content but require more energy in extraction and conversion of the raw materials to form the basic stampable sheet.  On a volume basis RIM and RRIM are by far the most energy efficient materials and processes. In addition, the cost for light duty RIM and RRIM equipment and their floor space requirements compare most favorably with the compression molding.

REFINING OF THE RRIM PROCESS, MATERIALS AND EQUIPMENT

     Our efforts were concentrated on a few basic materials which had the best heat stability.  In order to fine tune the RRIM materials and process to the required part performance - and also for future production - adjustments to the basic material composition, process and equipment were made.  Close cooperation among the material suppliers, custom molding company, and Ford Motor Company during many lengthy molding trials yielded an understanding of (A) material requirements, (B) equipment requirements, (C) process parameters, and (D) fibers.  We also established a meaningful RRIM part testing procedure.  The next important step in our development was to correlate the parts test results with the laboratory testing of the RRIM flat plaque.

Material Requirements (Table I)

     To thoroughly investigate the effects of part geometry and gate location on shrinkage and fiber orientation and the effect of fiber orientation on RRIM part heat stability, we conducted intensive studies in our processing laboratory.  All the measured parameters are being tabulated and complete results will be available in the near future.  The main areas we evaluated are:

     Heat Stability.  The heat stability of the RRIM part depends basically on:  proper selection of RIM material, process conditions, fiber shape and orientation, gate location, postcure temperature and time, part handling after molding.  In other words, the exterior RRIM part attached to steel car bodies must pass up to four times through the automotive assembly paint ovens with no difficulty.  The air temperature in the oven is 300°F.  Permanent cumulative expansion or contraction of the part must nor occur under these conditions.  The RRIM part must stay within the required assembly tolerances.  No visible distortion or sag are permissible.  This is the area where most of our work is now concentrated.  Study of fiber orientation and new high-heat stable materials is underway.

     Heat Sag.  The typical heat sag of unfilled heat stable materials varies from 0.1 in. to 0.3 in. and is measured on a four-inch cantilevered beam at 325°F/30 minutes.  If a distinction between similar materials must be made, it is advantageous to use an eight-inch cantilevered beam as well.  In order to understand the heat performance

Table I.  RRIM Material Properties

| MATERIAL NOMENCLATURE | REINFORCING AGENT | AUTOMOTIVE CLASS "A" FINISH (no ribs) | (with ribs) | MODIFIED GARDNER IMPACT (in lb.) | FAILURE MODE OF MODIFIED GARDNER IMPACT | CTE $1*/11**$ $(X10^{-6}$ In/ In/$^{\circ}$F) | FIBER ORIENTATION | FLEXURAL MODULUS $X10^3$ (PSI) |
|---|---|---|---|---|---|---|---|---|
| SEMI-HIGH MODULUS (RAPID) | A – None | Excellent | Excellent | 280 | Cracking | 80 | – | 60 |
|  | B – Reinforced 25% milled glass 1/16" | Excellent | Heat Sinks | 160 | Cracking | 23/38 | Yes | 180 |
| HIGH MODULUS | A – None | Excellent | Heat Sinks | 50 | Shattering | 68 | – | 100 |
|  | B – Reinforced 25% milled glass 1/16" | Excellent | Heat Sinks | 40 | Shattering | 20/35 | Yes | 250 |
| HIGH MODULUS (RAPID) | A – None | Excellent | Excellent | TBD*** | TBD*** | 69 | – | 100 |
|  | B – Reinforced 25% milled glass 1/16" | Excellent | Heat Sinks | TBD*** | TBD*** | 21/36 | Yes | 250 |
| SMC | B – 25-30% 1" chopped glass | Surface repair is required. | Surface repair is required. | 12 | Cracking | 10 | Negligible | 1200 |

\*    1 Parallel
\*\*   11 Perpendicular
\*\*\*  TBD – To Be Determined

of the RRIM material, one must measure the heat sag at the standard
automotive assembly paint oven temperatures:   (a) 325°F/30 min.
(taupe primer), (b) 250°F/30 min. (primer), (c) 300°F/30 min. (top
coat).

Fiber Orientation.  Parallel, perpendicular, and 45° orientations
of the fibers to the flow has been found in all the RRIM parts.   In
addition to the above-mentioned orientations, a phenomenon called
"fountaining" will occur when the liquid front of RRIM slurry splits
into several independent streams.  This phenomenon is a result of
abrupt change in liquid front direction and will affect the materials
CTE and shrinkage in that particular area of the RRIM part (Figure
1).   If fountaining occurs in a critical section of the RRIM part,
the parts performance will be negatively affected during heat expo-
sures.  To eliminate this unwanted effect, experimentation with gating
in the prototype tool is necessary before a decision on production
tooling is made.  Fountaining is not a characteristic of the RRIM flat
plaque (Figure 2).  Therefore physical properties obtained from flat
plaque and those of an actual part do not correlate.

Impact.  The fiber addition has a negative effect on the part
impact.  It was established that the impact strength is equal or better than SMC.  After extensive testing it was defined that not only
the actual impact value but also the failure mode of imparted sample
should be studied.  It was not possible to make a correlation between
the plaque impact and impact test of the part in car position.  This
is one of the areas where our work is being concentrated.

Class "A" Surface Finish and Paintability.  Past molding and
painting trials have proven conclusively that the surface quality of
RRIM parts designed without ribs retains the same excellency as an
unreinforced part.  The standard body enamel paint has been selected
in order that the parts may be painted at the same time as the steel
bodies, on the assembly line.  If the RRIM part needs to be designed
with the reinforcing ribs not hidden under a design line, the Class
"A" surface finish cannot be achieved with available knowledge and
techniques.  It is essential to the RRIM process that so-called

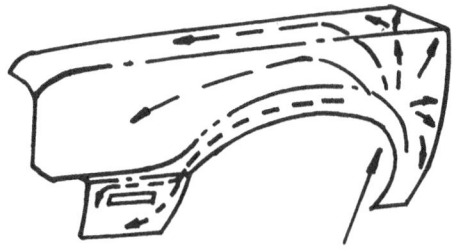

Figure 1.   Fountaining Effect in Automotive Part.

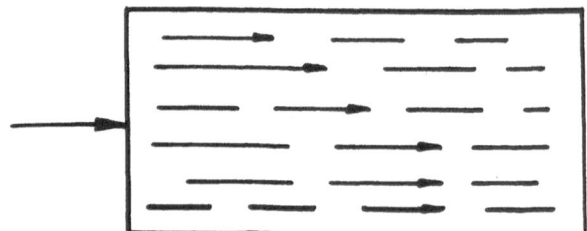

Figure 2.  Parallel Orientation in Flat Plaque.


"sink free" materials be on the high priority list at material development laboratories.

Flexural Modulus and CTE.  Flexural modulus should be identified
for a RRIM material, but the material system selection must be based
on material heat stability at 325°F, and impact strength at -20°F
rather than flexural modulus value.  The CTE plays a very important
role in the parts heat stability and a cumulative behavior for CTE
is not desirable (see heat stability earlier in this chapter).  The
CTE values fluctuate for unfilled polyurethanes between $50-90 \times 10^{-6}$
in/in°F.  The addition of 25% fibers will lower the CTE to $20-25 \times$
$10^{-6}$ in/in/°F.  The goal is to develop a material having CTE close
to aluminum $12 \times 10^{-6}$ in/in/°F.

Shrinkage.  It has been noted that each new part designed in
RRIM will have different and unique fiber orientation due to the position of the gating system and the shape of the part which translates
also into different material shrinkage.  This fact must be taken into
consideration when designing RRIM parts.  Precise prototyping and
testing of the actual RRIM part to allow the molder and moldmaker to
compensate for the shrinkage factor is essential.

Equipment Requirements

Versatility of the RRIM metering and mixing machines becomes
more important as we continue in the development of the RRIM technology.  The Universal RRIM machine of the future must be sensitive
to the changes in the ever-expanding field of RRIM materials and
parts.  As we see it at Ford Motor Company, the Universal RRIM machine
should be capable of accepting rapid changes in:

Shot Size (Retrofitting).  Consideration is given to the parts
ranging in weight between 2-50 lbs.  This broad weight scale of RRIM
parts will necessitate retrofitting capability in metering cylinders
or pumps, so that all part sizes can be accommodated.

Material Ratio (Retrofitting).  The mixing ratio of future materials will be different due to the new demands on the material's phys-

ical properties.  The material proportions of the RRIM systems will range anywhere from 0.5:1 to 1:1 to 1:3 to 1:0.5 to 3:1.  We must be prepared to properly meter and mix these new evolving materials with the RRIM equipment.

Machine Output.  New rapid materials are being introduced on the market.  These materials are capable of fulfilling the requirements of heat stability and impact.  (Checked on plaques and small parts 1-3 lbs.).  The difficulties occur when molding of parts larger than 8 lbs. is required.  The rapid material solidifies before it has time to flow and fill out the entire mold.  It is essential to the future of RRIM that the machines be rated not only on output of 1-7 lbs/ second but also for an output of 10-20 lbs/second.  The Universal machine can be capable of molding:  small parts using standard materials, small parts using rapid materials, large parts using standard materials, large parts using rapid materials.  For the machine, a combination of large hydraulic capacity and the retrofitting feature is essential for the above.  There must be much more work done in the area of rapid materials for large exterior body parts, but equipment development is fundamental to their future.

## Process Parameters

There is a clear need for higher productivity in the RRIM processing.  Polymerization time of the rapid RIM materials is as short as 15 seconds, yet mold cleaning extends the overall cycle time to 2-3 minutes. The addition of proven internal mold release would significantly increase the productivity of the RRIM process.

A type of "In-Mold Coating" spray (IMC) is being investigated as one of the solutions to the required higher productivity objective. This technique would eliminate the priming of the part at the manufacturing plant.  The work is in progress and information will be available shortly.

## Fibers

Fiber Addition.  Utilization of fibers will improve the CTE of the RRIM materials and will also substantially increase their heat stability.  Both are needed for proper functioning of the RRIM part assembled on the steel car body.  The addition of fibers will negatively affect the elongation of the material.  In order to make use of the positive effect of fiber addition, a search for materials with very high elongation combined with high heat stability is underway.

An ideal material for exterior body panels would be as follows:

| RRIM Material Property | Requirement |
|---|---|
| 1. Elongation (filled or unfilled) | 350–400% |
| 2. Heat stability of the part in car position (filled or unfilled urethane or non-urethane material | 400°F or 325°F |
| 3. Complete material cure | Required (to be specified by material supplier) |
| 4. Permanent cumulative expansion or contraction during four (4) repeated heat cycles (285°F–300°F) | Must not occur |
| 5. CTE | $12 \times 10^{-6}$ in/in/°F or lower |
| 6. Specific gravity (unfilled material) | 62 lbs/$FT^2$ or lower |

Fiber Shape, Length and Its Distribution. During intensive study in our processing laboratory, it has been demonstrated that the fiber shape, length and its distribution plays an important role in the development of a RRIM part. The late phenomena and their effect on part performance will be studied more thoroughly in order to determine the best type of filler for RRIM technology. It is likely that various shapes of fillers will be used for different RRIM applications due to varying orientation characteristics. The selected fiber must help to improve the physical properties and performance of the automotive part; it cannot be used only as an agent lowering the price of RRIM material.


NEW DEVELOPMENTS

A brief update on new developments in the non-polyurethane field will be given with no intention to classify the materials for particular automotive applications. The field is wide open and some of the new materials are still in laboratory test tubes. This is an area where we can expect the most surprising developments. This section discusses:

## Materials

The need to handle the RRIM automotive exterior parts with steel car bodies at 400°F dictates investigation of materials other than polyurethanes. The non-polyurethane RRIM technology requires: low viscosity liquid materials (50-2,000 cps at 72°F), fast reacting materials (15-60 seconds), compounded materials systems with internal mold release, molded material capable of Class "A" surface finish.

Some of the physical properties of non-polyurethane materials are identified in Table II. The heat sag properties of glass mat-reinforced epoxy tested at 400°F are encouraging. The impact strength of the same material is better than SMC.

The combination of a RIM material reinforced with fiber mat should be investigated very closely; under hood components and possibly some structural members are likely candidates. Techniques for holding the fiber mat in place during injection of RIM material must be studied.

In order to understand the behavior of the new materials under environmental conditions, it is important to test their water absorption characteristics over a period of at least ten (10) days.

## Equipment

It is desirable to develop RRIM equipment capable of processing polyurethanes and non-polyurethanes (epoxies, polyvinylesters and others) rather than individual equipment for processing each of these materials. This is very important to the future of the RRIM process in general and should be considered while it is still in its early stages of development.

There are some materials which cannot be processed with today's RRIM equipment. In these cases equipment development needs to proceed closely with the material development. This must be done in order to bring the new materials to the point where they can be tested in the automotive processing laboratories.

## Chopped Fibers

Limited quantities of chopped fibers supplied as separate filament rather than bundles have been tested at Ford Motor Company. It has been observed that material filled with 15% chopped fibers is equal in stiffness to the material filled with 25% milled glass. This is encouraging because the weight of the part will decrease.

There are two basic questions on chopped fiber behavior in automotive parts which need to be answered: (1) how the chopped fibers influence shrinkage, and, in particular, uniformity of shrinkage,

Table II.  Physical Properties Values of Non-Polyurethane Materials

| MATERIAL | REINFORCEMENT | FLEX MODULUS (PSI) X 10³ | HEAT SAG (in) | ELONGATION (%) | MODIFIED GARDNER IMPACT (in lb) | FAILURE MODE OF IMPACTED SAMPLE | WATER ABSORPTION (%) |
|---|---|---|---|---|---|---|---|
| SMC | 25-30% (1" chopped fibers) | 1200 | Negligible | 1.5 | 12 | Hairline Cracks | 0.1 (24 hr.) |
| LIQUID EPOXY I | None | 400 | 325°F .6 | 10 | 20 | Shattered | TBD |
| LIQUID EPOXY II | 10% milled glass 1/8" | 500 | 325°F 0.07 | 5 | 10 | Shattered | TBD |
| LIQUID EPOXY III | Single thickness glass mat (2 oz.) | 650 | 400°F 0.009 | 3 | Tested only to 30 in/lb | Hairline Cracks | TBD |
| LIQUID EPOXY IV | Double thickness glass mat (2 oz.) | 960 | 400°F 0.006 | 3 | Tested only to 30 in/lb | Hairline Cracks | TBD |
| LIQUID NYLON | None | 28 | 1.17 | 32 | 240 | Small Indentation | 7.6 (10 days) |

(2) the influence of the chopped fibers and their orientation on the heat stability of the part.


CONCLUSIONS

The first successful steps in developing and establishing the basic RRIM concepts through machinery builders laboratories, custom molders, chemical companies and the automotive industry have been achieved.

The RRIM technology is coming out of its infancy; however, much more work is required to refine the equipment, the process and especially the materials.

We believe that most of the innovations in the near future will be accomplished in the area of RRIM technology, either in polyurethanes or non-polyurethanes.

Intensive industry efforts are needed in: developing of RIM materials with heat stability at 400°F combined with good impact characteristics, resolving of the fiber orientation and its influence on part performance and shrinkage, developing of sink free materials, designing and testing of the bulk fiber handling system, developing techniques for holding the fiber mat in place during injection of RIM materials.

# METERING AND MIXING OF RIM REACTANTS

Fritz W. Schneider

pta martin sweets Co.
P. O. Box 1068
Louisville, KY 40201

## INTRODUCTION

The objective of a high pressure metering and mixing system in accordance to the RIM principle is to meter the liquid reactants, in general polyol and isocyanate, in the required volumes and ratios and mix them efficiently.

The dispensed material may be, depending on the application, directed in an open mold, in a closed mold through a porthole or by way of a channel, aftermixer device and film gate (Figure 1). The expression RIM, meaning "Reaction Injection Molding," is not completely correct for open pour fill. However, with the understanding that RIM can be interpreted also as "Reaction Impingement Mixing," taking reference to the mixing principle, it may be used as a general term.

The final product manufactured on RIM equipment may be as different as an insulation panel, furniture parts, skis, cushions, automobile bumpers and shoe soles (Figure 2). A RIM metering and mixing system consists basically of the material tanks for the reactants connected by a recirculation pipe system to the metering pumps and the impingement mixing head (Figure 3). Heating and cooling units are required for temperature control of the reactants as well as a hydraulic unit for the operation of the mixing head.

The metering pumps not only supply the materials in the required volume, but also the mixing energy in the way of pressure. High pressure operation in the system is necessary before, during and after a dispense cycle for insuring uniform conditions from the very beginning to the end of dispense.

243

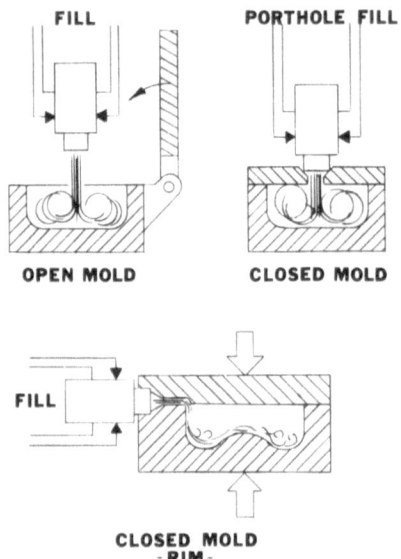

Figure 1.   Applications for high pressure impingement mixing heads:
            open mold fill, closed mold porthole fill, closed mold
            RIM via aftermixer device and gates.

     Key element of the metering and mixing system is the impingement
mixing head featuring a hydraulically operated plunger in the mixing
chamber for controlled injection of the reactants as well as for solvent free cleaning after the dispense operation (Figure 4).

     The main objective of this paper is to discuss the principle of
impingement mixing for better understanding.

MAJOR FACTORS AND REQUIREMENTS FOR EFFICIENT IMPINGEMENT MIXING
OF REACTANTS

     The mixing head has to produce a well mixed material with high
efficiency in regard to energy consumption.

     Figure 5 shows schematically a cross section through an impinge-
ment mixing device with retracted cleaning plunger.  The reactants
are supplied by the metering pumps to the impingement nozzles with
relatively high pressures of 1500 to 3000 psi.  There, the pressure
energy is transformed into energy of velocity.  The flow speeds in
the nozzles are in the range of 300 to 500 ft. per second.  The pressure in the mixing chamber is close to zero for open mold pouring and
usually not exceeding 100 to 150 psi for RIM applications.

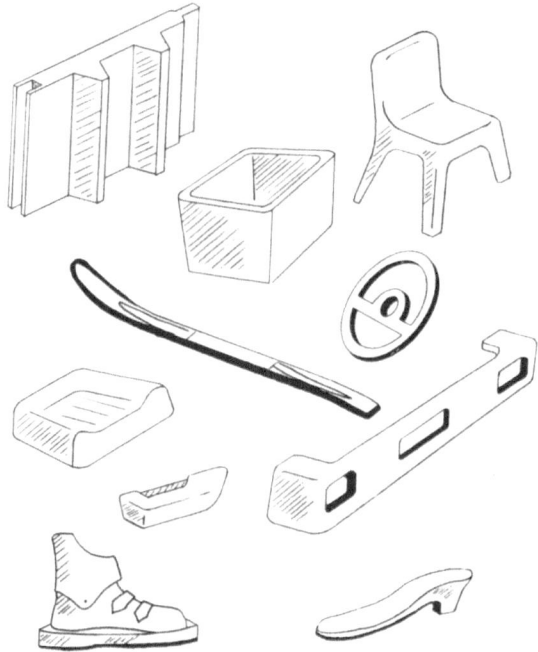

Figure 2.   Examples of products manufactured with high pressure RIM
            equipment.


For the achievement of a good mix, the flow conditions in the
mixing chamber have to be turbulent.  A deciding factor for the prediction of turbulence is the critical Reynolds' number of a mixing
system, marking the transition between laminar and turbulent flow
conditions.

The actual Reynolds' number at which a system has to be operated
has to be larger than the critical Reynolds number:

Re laminar  <  Re critical  <   Re turbulent

The Reynolds' numbers can be calculated at knowledge of the velocities of the material streams entering the mixing chamber, their
viscosities and densities and the geometrical configurations of the
nozzles.


THROUGHPUT CAPACITY OF NOZZLES

The throughput capacity of an injection nozzle can be described
by the following term:

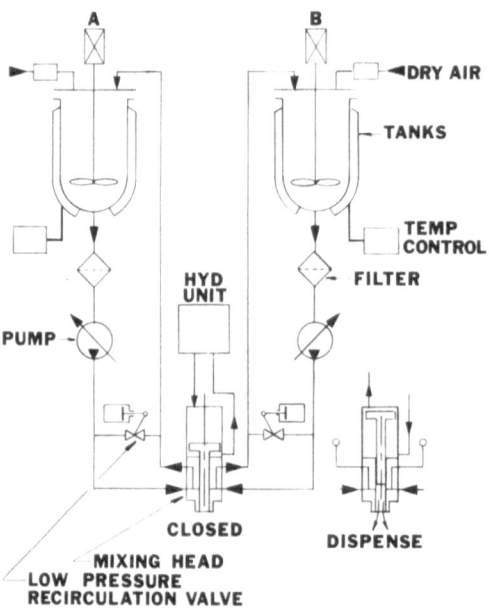

Figure 3.  Schematic flow diagram of a high pressure RIM metering and
dispensing machine with self-cleaning impingement mixing
head.

$$Q = v \cdot \alpha A \cdot \rho \qquad\qquad\qquad\qquad\qquad\qquad \text{kg/sec} \qquad (1)$$

$v$ = velocity of flow                                                      m/sec
$\alpha$ = contraction factor in regard to nozzle
    design
$A$ = area of nozzle orifice                                               $m^2$
$\rho$ = density of liquid                                                 $kg/m^3$

The average velocity at the outlet of the nozzle is in accordance to the law of Bernoulli:

$$v = \beta \sqrt{2\, g\, \frac{\Delta p}{\rho}} \qquad\qquad\qquad\qquad\qquad \text{m/sec} \qquad (2)$$

$\beta$ = friction coefficient
$g$ = earth gravity = 9.81                                                 $m/sec^2$
$\Delta p$ = pressure difference = $p_1 - p_2$                             $kg/m^2$

The capacity relationship for a nozzle with circular orifice
is in accordance to (1) and (2):

$$Q = \frac{d^2 \cdot \pi}{4} \cdot \rho \cdot k \sqrt{2\, g\, \frac{\Delta p}{\rho}} \qquad\qquad\qquad \text{kg/sec} \qquad (3)$$

d = diameter of nozzle orifice
k = α • β = nozzle factor = 0.7 to 0.8

For better convenience of calculations, formula (3) may be written also as:

$$Q = \frac{d^2 \cdot \pi}{4} \cdot \rho \cdot k \sqrt{20g \cdot \frac{\Delta p}{\rho}} \qquad \text{g/sec} \qquad (4)$$

by putting in the dimensions for:

d in mm            g = 9.81            m/sec$^2$
$\rho$ in g/cm$^3$            p in kg/cm$^2$

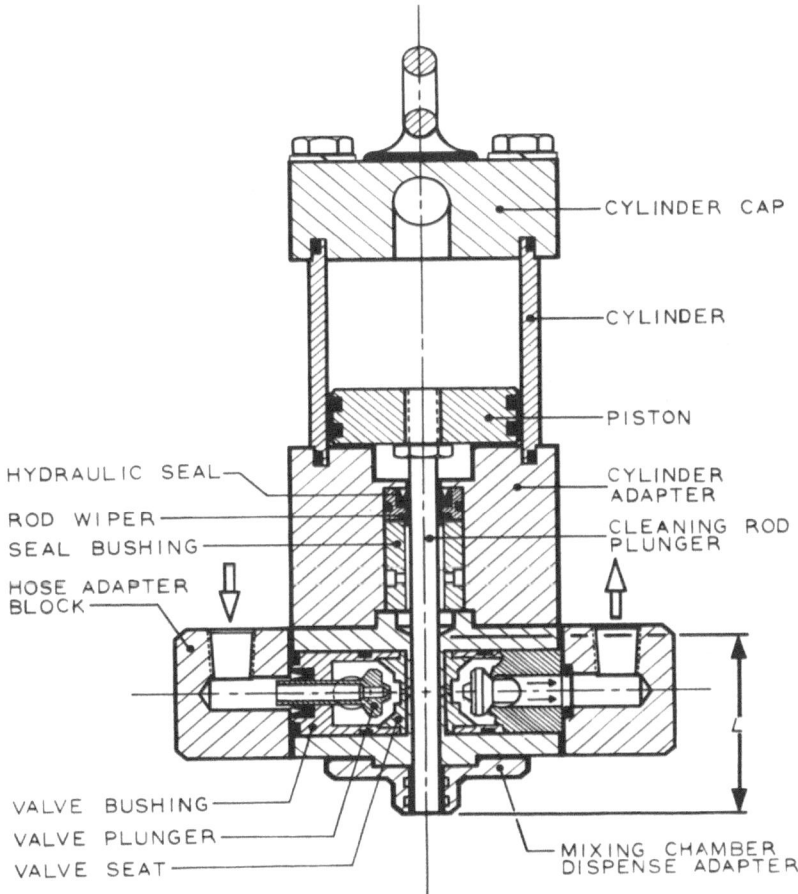

Figure 4.   Cross section of a high pressure impingement mixing head for free pour and RIM applications.

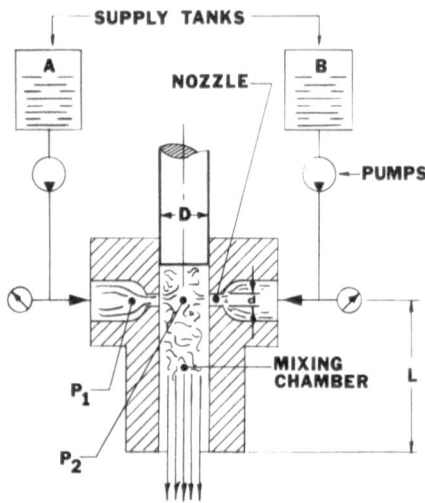

Figure 5.  Schematic cross section through a high pressure impingement mixing head in mixing and dispense mode.


VISCOSITY

The viscosity of the chemicals, as well as the final viscosity
of the mixture, before reaction takes effect, has a major influence
on impingement mixing efficiency.

The dynamic viscosities of liquids are usually measured and num-
bered in Poise (P) or centi Poise (cP).  The following relations
exist:

Dynamic Viscosity

$$\mu = 1 \text{ cP} = 1 \cdot 10^{-2} \text{ Poise (P)} = 1 \cdot 10^{-3} \quad \text{kg/sec m} \quad (5)$$

For further consideration, the <u>kinematic viscosity</u> is helpful:

$$\nu = \mu/\rho \qquad\qquad\qquad\qquad \text{m}^2/\text{sec} \quad (6)$$

$$\rho = \text{density} \qquad\qquad\qquad\qquad \text{kg/m}^3$$

The kinematic viscosity indicates the ratio of shearing stress
to shearing strain in a liquid.

REYNOLDS' NUMBER (Re)

The Re-number is dimensionless. It describes the ratio of in-ertia forces to viscous forces acting in the accelerated flow stream. In operating a system at larger Re-numbers as Re critical, the in-ertia forces are in excess of the viscous forces and the system runs turbulent.

The Reynolds' number for a circular flow pattern can be calcu-lated by the following terms:

$$Re = d \cdot v / \nu = d \cdot v \cdot \rho / \mu = 4 \cdot V \cdot \rho / \pi \cdot d \cdot \mu = 4Q / \pi \cdot d \cdot \mu \qquad (7)$$

or, in regard to formula (4) for convenience:

$$Re = \frac{d \cdot \rho \cdot k}{\mu} \sqrt{20g \cdot \frac{\Delta p}{\rho}} \cdot 10^3 \qquad (8)$$

d = diameter of nozzles orifice     mm
$\rho$ = density     $g/cm^3$
g = earth gravity = 9.81     $m/sec^2$
p = pressure difference     $kg/cm^2$
$\mu$ = dynamic viscosity     cP

An ideal mixture is achieved if the single elementary particles of the liquids to be mixed are meeting with each other individually, after the breaking up of the viscous forces which were holding to-gether the particles originally by the inertia forces charged to the particles by transformation of pressure energy to dynamic energy of velocity (Figure 6).

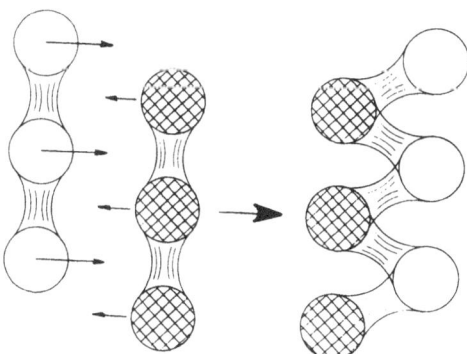

Figure 6. Schematic of impingement mixing.

The transformation of pressure energy to mixing energy is complete, if full turbulent conditions in the mixing chamber are achieved. However, the mixed materials are leaving the dispense opening again under laminar flow conditions without stationary turbulence. Under ideal conditions, the pressure energy is totally transformed to heat energy. The temperature increase of the mixed materials is for example 7°C or approximately 12°F at a pressure drop of 150 kg/cm$^2$ and an assumed specific heat capacity of c = 0.5 kcal/kg°C (Figure 7).

Basic research of mixing parameters and mixing efficiency was executed by different methods, as for example, by employing non-reacting color dyed sample liquids as polyols and water/glycerine solutions as substitute for isocyanates.

Mixing efficiency and quality were analyzed with similar methods as employed to measure the quality of emulsions.

Another method employs the adiabatic temperature rise over time, achieved by the exothermic reaction of the mixed chemicals (Richter

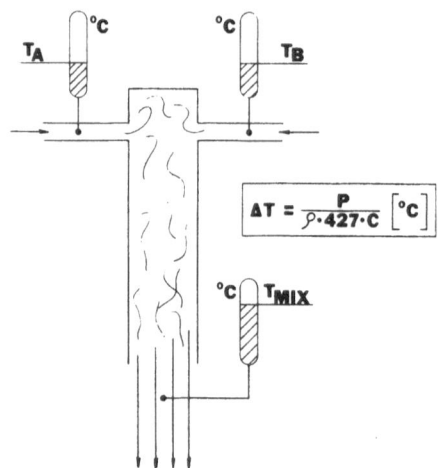

Figure 7.   Temperature increase by energy transformation:

Transformed pressure:   $P = \dfrac{Q_A \times \Delta P_A + Q_B \times \Delta P_B}{Q_A + Q_B}$   kg/m$^2$

Density:  $\rho = $ kg/m$^3$

Mechanical thermo equivalent:   1kcal = 427 mkg

Specific heat capacity:   $c = \dfrac{Q_A \times c_A + Q_B \times c_B}{Q_A + Q_Q}$   kcal/kg°C

In the figure: $\Delta T = \dfrac{P}{\rho \cdot 427 \cdot C}$ [°C]

and Macosko, 1978; Lee and Macosko, 1978).

By analyzing tests results and practical applications, it was
found that for the achievement of turbulent mixing conditions, impingement mixing devices have to be operated by Reynolds numbers of:

Re > Re Crit. = 50 to 150 (200)

Lower Re numbers are required, and for that reason, less energy,
for mixing heads featuring circular fixed nozzles correctly aligned
against each other, higher Re numbers for devices where the nozzles
are arranged in an angle or entering the mixing chamber tangentially
(Figure 8).

CONDITIONS AT DIFFERENT RATIOS OF CHEMICAL REACTANTS

As a general rule, both material streams impinged in the mixing
chamber should meet the criteria for turbulent mixing conditions.  In
addition, the main mixing area should be kept, if possible, in the
center of the mixing chamber.  This can be achieved by pressure adjustments in order to receive balanced energy momentums of the mass

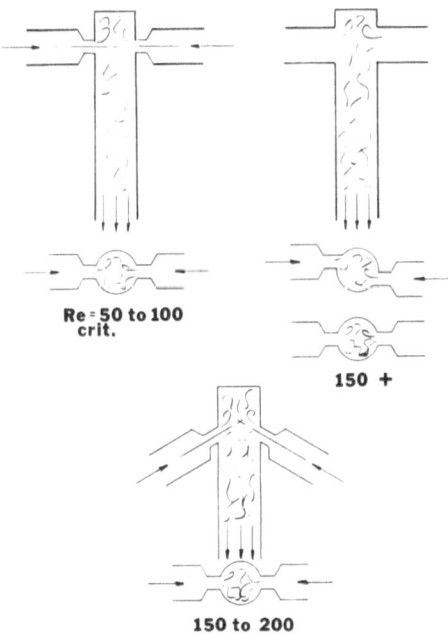

Figure 8.   Critical Reynolds' numbers of impingement mixing devices
            of different design and nozzle arrangements.

flow streams entering the mixing chamber in accordance to the following term:

$$Q_A \cdot \Delta p_A = Q_B \cdot \Delta p_B \tag{9}$$

$Q_A$, $Q_B$ = capacities of Component A and B

$\Delta p_A$, $\Delta p_B$ = pressures of Component A and B

This relationship can usually be kept closely for ratios of the
chemicals A:B = 1:1 to 1:1.5 or vice versa. For higher ratios, com-
promises have to be made, because of pressure requirements exceeding
the maximum operation pressures of metering systems. Fortunately,
the loss in mixing efficiency, mainly because of increased friction
on the wall of the mixing chamber, at conditions not completely in
line with term (9), is gradual. In practice, RIM formulations with
A:B = 1:4 are successfully processed. For materials with a longer
pot life and some affinity of the components to each other, as for
example coating materials, epoxy formulations, etc., it was exper-
ienced that successful processing was possible at ratios of up to
1:10. Research is not completed in this respect.

## ADDING OF FURTHER FLUID COMPONENTS TO THE MIXING CHAMBER

Components other than the main reactants may be impinged in
the mixing chamber in addition, with satisfactory results, as long
as several conditions are met and certain limitations observed.
Fluid additives may be: air, blowing agent, color paste, catalyst,
flame retardants, etc.

In general, the additives should be directed in the mixing chamber in an area and way that they do not disturb the turbulent mixing
conditions created by the main streams. In case the additives have
viscosities equal or higher than those of the main reactants, it is
difficult to achieve turbulent conditions in the additive stream itself due to their usually small mass flow, depending on ratio relationships. The turbulence created by the main streams have to provide uniform distribution. A higher input of mixing energy may be
necessary.

## MIXING OF REACTANTS WITH SUSPENDED SOLID FILLERS IN ACCORDANCE
## TO RRIM

The general rules to achieve good mixtures for reactants, in
which relatively coarse solid filler particles are suspended (Reinforced Reaction Injection Molding), are the same as for non-filled
materials.

The most common filler in practical use consists of milled glass fibers. Other fillers tested include:  barium sulfate, calcium carbonate, ground rubber, saw dust.

The fluid characteristics of solid filled polyurethane systems show a non-newtonian behavior.  At the high shear rates of the materials passing through the injection nozzles with high velocities, the high apparent viscosities of the slurries at rest drop dramatically and approach the values of the unfilled chemicals.

By practical experience, it can be stated that probably any material slurries can be mixed, presupposing that they can be supplied to the mixing chamber at the required velocities without clogging of the nozzles.  In practical tests, it was surprising that parts manufactured with filled materials display often less flow patterns and better surface quality than parts made of non-filled but otherwise equal chemicals and similar process conditions.  The explanation may be that the suspended particles, each having a certain mass and inertia, have a movement of their own in the liquid providing by that mixing assistance and compensating for the influence of higher viscosity.  In addition, the solids act favorably as mechanical stabilizers of the initial mixture, which is basically an emulsion, until the increase of viscosity at progressing reaction prevents possible separation of liquid components.  Further research of this phenomenon would be highly interesting and would lead probably to interesting conclusions and applications.


LEAD LAG PHENOMENA

The objective of a metering and mixing system is to provide a good uniform mixture of the reactants at exact stoichiometric ratios from the very beginning to the end of the dispense operation.  However, this cannot always be obtained.  Many operations suffer from so-called "lead lag" problems.

The term "lead lag" describes the problems initiated by non-uniform or off ratio flows of the reactants at the instant of starting or stopping the dispense in "on-off" mixing devices.  One chemical may lead or lag in regard to the other if no complete control of the flow conditions is obtained.  The results of "lead lag" material entering the mold may be the production of unacceptable quality.

A major cause for lead lag problems are sudden pressure changes in the line systems at the opening and closing of the mixing head. All mixing devices feature, in one way or the other, external or internal valve actions directing the continuous flow of the metering pumps from high pressure recirculation to impingement in the mixing chamber.  In most mixing devices, the flow streams are shut off

briefly which leads inevitably to accumulations of the materials in
the hose and line systems acting like spring loaded accumulators.
Designs, in which at opening of the mixing head the switch-over occurs
on stream without pressure spikes, have an advantage.  Otherwise, the
arrangement of "lead lag" collecting chambers in aftermixer devices
may be helpful, for catching the first incoming material in dead end
channels (Figure 9).

BASIC REQUIREMENTS FOR THE REACTANTS FOR RIM PROCESSING

Reactants mixing well when hand mixed or processed on low pressure machines are not necessarily RIM materials.

For the development of materials suitable for the RIM process,
it has to be considered that the mixing time in an impingement device
is only a fraction of the time in an impeller mixer.  The viscosities
have a major influence for turbulent conditions, necessary for a good
mix.  The temperature of the mixture is increased, equivalent to the
higher energy input.

Special attention has to be spent for preventing separation effects of an initial good mix after dispense in the mold, until reaction and gelling is so far progressed that the mixture is stabilized.
The problems involved are complex and often difficult to analyze in

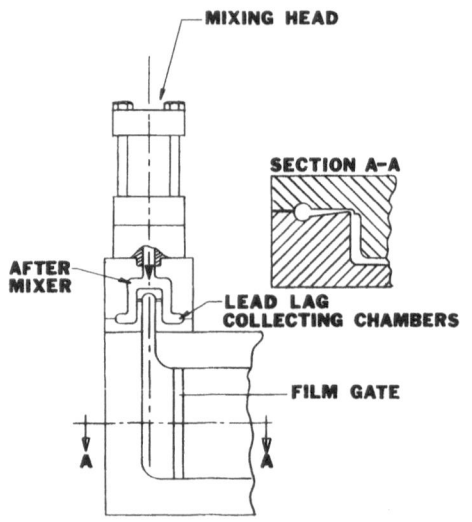

Figure 9.  Typical arrangement for mixing head, aftermixer device
runner and filmgate on RIM mold.  Cavities for aftermixer and gate system are in the parting line of the
mold for easy demolding.

the short span of time they occur.  They may relate to the progress
of reaction, presence of intermediary reaction products, affinity of
the products to each other and rheology of the mixture.  The problems
can be approached by the employment of aftermixer devices for pro-
longation of the mixing time, additives acting as emulsifiers and
stabilizers as well as in the case of foam products, addition of gas
(air) for increased nucleation and stabilization of the mixture due
to the thixotropic behavior of gas loaded material.

The complex nature of the RIM process makes it necessary that
the development and evaluation of RIM materials is executed under
realistic conditions.

For operation with small quantities several laboratory machines
are available.  Typical examples are the "Mini-RIM," operating with
metering pistons and the "HP 20L," an industrial production type machine modified for laboratory employment.

The "Mini-RIM" - metering and mixing machine was designed from
the beginning as a laboratory unit for RIM material development,
evaluation of materials, basic fluid mechanic research, as well as
for mold gate and aftermixer studies.  The original design is based
on a development of a laboratory research unit by the University of
Minnesota.

The machine was redesigned and modified for meeting industrial
R/D requirements.  Today's improved design represents a very flexible
laboratory unit capable of processing RIM formulations of different
types.  Only small quantities of chemicals are required to produce
test samples.  The machine features simple design and easy time-sav-
ing clean out for change of materials.

Besides polyurethanes, the machine processes epoxy resins, unsaturated polyethers, silicones, nylons, acrylics, etc.

The operation of the "Mini-RIM" with filled materials is possible on a limited basis, as long as the materials can be passed,
without clogging, through the relatively small nozzle orifices and
the line system, restricted in size by this type of machine.

The "Mini-RIM" is a true, small size RIM machine featuring
countercurrent high pressure impingement of the chemicals in a mixing head having a cylindrical mixing chamber and mechanical, solvent
free cleaning by a hydraulically operated cleaning rod.  The mixing
head employed is a "pta martin sweets" type "A" head with a mixing
chamber diameter of 0.250 inch.  This head is also used for small
production machines.  The main feature of the mixing head is a self-
controlled internal valving system, allowing switch-over from high
pressure recirculation to dispense without pressure spikes in the
line system reducing lead lag phenomena and mechanical fatigue.

(For more mixing head information see Urethanews "pta martin sweets"
High Pressure RIM Impingement Head.)

The "Mini-RIM" system is rated for injection pressures of up to
3000 psi as employed also in the larger industrial production machines. The machine features high pressure recirculation through the
mixing head before and after the dispense operation for uniform mixing pressure conditions from beginning to end of dispense and for the
clearing of the entrance gates to the mixing chamber from reactive
material.

The nozzles have circular orifices of fixed diameter for high
efficiency and less sensitiveness for clogging. They are quickly
interchangeable to adapt for differing requirements of capacity and
impingement pressure.

The maximum shot volume of the metering system is approximately
300 $cm^3$ at volumetric ratio number of 1:1. By adjustments of the
stroke ratio of the metering cylinders and exchange of the cylinders
of different size of diameter, the ratio A:B can be varied from 0.8:1
to 3.1:1.

The exact metering is executed by two mechanically linked cylinders actuated by a common hydraulically operated drive cylinder.

Accurate ratio calibration of the system is executed under actual shot conditions by the employment of a calibration adapter to
the mixing head. The chemicals injected simultaneously in the mixing
chamber, if the mixing head is opened, are caught in grooves and
channelled out separately for the checking of shot weight and ratio.

The shot size and volume of the material tanks are optional.
The smallest recommended tanks have a capacity of one (1) quart.

Figure 10 shows a schematic flow diagram of the machine.

The "Mini-RIM" unit consists basically of the metering system,
mixing head, line and valving system, supply tanks, hydraulic drive
system, temperature control and electric control system.

For convenience, the metering and mixing unit is installed on
a mobile table. The metering cylinders have the diameters $d_1$ and
$d_2$. One cylinder is directly linked to the hydraulic drive cylinder.
The second cylinder, arranged parallel to the first at the distance
"A", is linked to the drive by means of a crossbar. The crossbar
swivels around an adjustable pin point located at the parallel distance "A + B" from the joint of the drive piston.

The volumetric capacity ratio is in accordance to the geometric
relations:

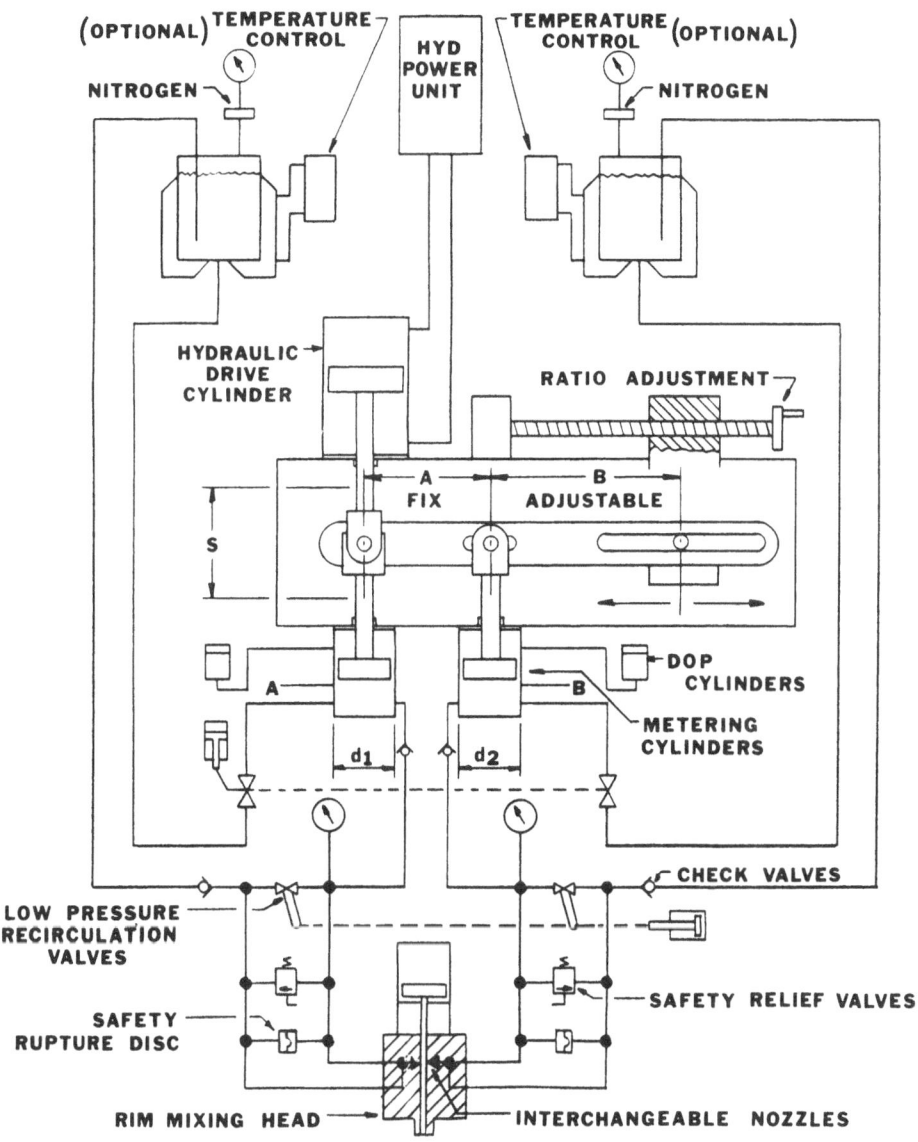

Figure 10.   Schematic flow diagram of the "pta martin sweets" MINI-RIM high pressure metering and mixing machine, HP-6AC.

$$\frac{V1}{V2} = \frac{A + B}{B} \cdot \frac{d_1{}^2}{d_2{}^2}$$

The weight capacity ratio is:

$$\frac{Q_1}{Q_2} = \frac{A + B}{B} \cdot \frac{d_1{}^2}{d_2{}^2} \cdot \frac{\rho_1}{\rho_2}$$

$d_1$ ; $d_2$ = diameters of metering cylinders

$\rho_1$ ; $\rho_2$ = densities

The length "A" is a fixed dimension. The adjustment of "B" is executed by a manually operated spindle and slide device with scale indication.

The drive cylinder, metering cylinders and mechanical ratio adjustments are built up on a common base structure.

For low pressure recirculation of the materials, for flushing of the lines and temperature conditioning of the system, the drive cylinder is put in back and forth oscillation at full stroke, by that actuating, also, the metering pistons.

At the back stroke movement, adjusted for slow speed to insure a good fill, the chemicals are transported, by suction, from the material tanks to the metering cylinders by way of the feed lines and the open, pneumatically controlled feed line valves. The tanks may be slightly pressurized by nitrogen or dry air blanket on top of the materials, for compensation of pressure losses in the lines by friction due to the often highly viscous materials and preventing cavitation and development of gas bubbles in the lines and cylinders. Tank sizes are optional. Smallest tank volume is one-quarter gallon.

The low pressure recirculation valves, actuated pneumatically, are open in this mode. Check valves are arranged in the return lines for preventing suction through the return lines.

At final backward position of the drive piston, the direction of the oscillating movement is automatically reversed, by impulse from a limit switch, for delivery stroke. Simultaneously, the feed line valves are closed. By that, the materials in the metering cylinders are forced out. The major part of the materials at low pressure recirculation takes the way to the return lines through the low pressure recirculation valves. However, a small portion is also forced through the mixing head, flushing, by that, the nozzles and the lines connected to the head.

For precise temperature control, the materials may be directed through heat exchangers of the temperature control units, or the material tanks may have jackets which are supplied with media for temperature conditioning.

The machine is now ready for calibration and adjustments after constant preset temperatures are achieved.

The nozzle sizes are selected in accordance to capacity and pressure requirements, with the assistance of a diagram. The approximate ratio is selected by means of the ratio adjustment device, the total capacity by the adjustment of the hydraulic pump supplying the media for the main drive cylinder.

To put the machine in high pressure recirculation mode, the low pressure recirculation valves are closed.

For checking the ratio and final adjustments, the calibration adapter may be applied to the dispense end of the mixing head, allowing to collect the chemicals unmixed and separately as they leave the mixing head under actual dispense conditions when the cleaning plunger is retracted.

The ratio is corrected by turning the ratio adjustment device. The required shot volume is set by the adjustment of the effective stroke length "S" of the drive piston shaft by a digital shot counter. Pressures may be adjusted by varying the volume of the hydraulic drive system or exchange of volumes.

At the forward stroke of the metering piston, the displaced materials are forced in total through the mixing head and the injection nozzles. The pressure build-up is instantaneous; however, a short length of stroke is required to achieve uniform pressures.

If a dispense shot is called off by putting the controls in shot mode, the mixing head opens automatically after full pressure build-up. The materials are injected with high speed in the mixing chamber, mixed and dispensed at the front end. The digital shot counter gives the signal to push the cleaning plunger in forward position again and, by that, ending the shot. The rest of the volume, in regard to the total stroke length of the drive piston, is employed for post high pressure recirculation of the materials through the head for the flushing of the entrance ports to the mixing chamber and for preparation for the next shot.

For the preparation of test samples, an easy to install plaque mold with built-in aftermixer device, mold gate and electric heating may be applied to the front end of the mixing head.

Change of materials and flushing and cleaning of the lines can
be executed efficiently and relatively quickly with a minimum loss of
chemicals and solvents due to the small tank size and line system.
The procedure is to drain the chemicals out of the system, fill up,
to some degree, with adequate solvents, as for example D.O.P, and op-
erate the machine for some time in low pressure and high pressure
mode. Then drain again. Repeat flushing if necessary.

Another metering and mixing machine for R/D is the HP-20L. The
machine is basically a small production unit with a capacity of appr.
5 to 20 lb/min based on a proportional mix ratio of 1:1. It features
the same design principles as larger machines.

Figure 11 displays the flow diagram of this machine with material
tanks, continuously operated metering pumps, recirculation lines to
the mixing head and back to the tanks, as well as bypass lines with
shut off valves for low pressure and high pressure recirculation
through the mixing head. Temperature control is achieved by heat exchangers in the return lines and sensor controlled heater/cooler systems. Suggested tank size is 2 gallons. The tanks are equipped with
agitators, level control and means for nitrogen or dry air blanket
and prepressure on top of the materials.

The metering pumps, protected by edge type filters are of axial
piston type. Accurate flow control is achieved by speed control of
the pumps by D.C. drive motors.

The mixing head is a size "A" or "A/2" RIM mixing head of pa-
tented design with mechanical cleanout of the mixing chamber. The
"A/2" head having a diameter of the mixing chamber of 1/8" allows
flow rates as low as 30 to 40 gr/sec.without pressure spikes in the
system or significant lead lag. The machine is designed for a maximum operating pressure of 3,000 P.S.I. A hydraulic unit is included
for the operation of the mixing head. For prevention of accidental
cross over the unit is equipped with pressure switches which immediately close the mixing head if pressures are below minimum impingement pressure as well as pump shut down if the system is overpres-
sured. This can be caused by nozzle clogging or misadjustment of
pumps. As a final safety device against overpressure, rupture discs
are employed. Calibration may be executed by a through the head device, providing the most accurate method under actual dispense conditions, as well as by simultaneously operated 3-way valves in the
return line.

The mixing head can be attached fixed to a mold in accordance
to the RIM principle but also employed for open mold fill. Depending
on the chemical system employed, the machine may be capable to pour
shots of a few grams up to several pounds.

The "HP 20L" is the more verstaile unit allowing larger varia-

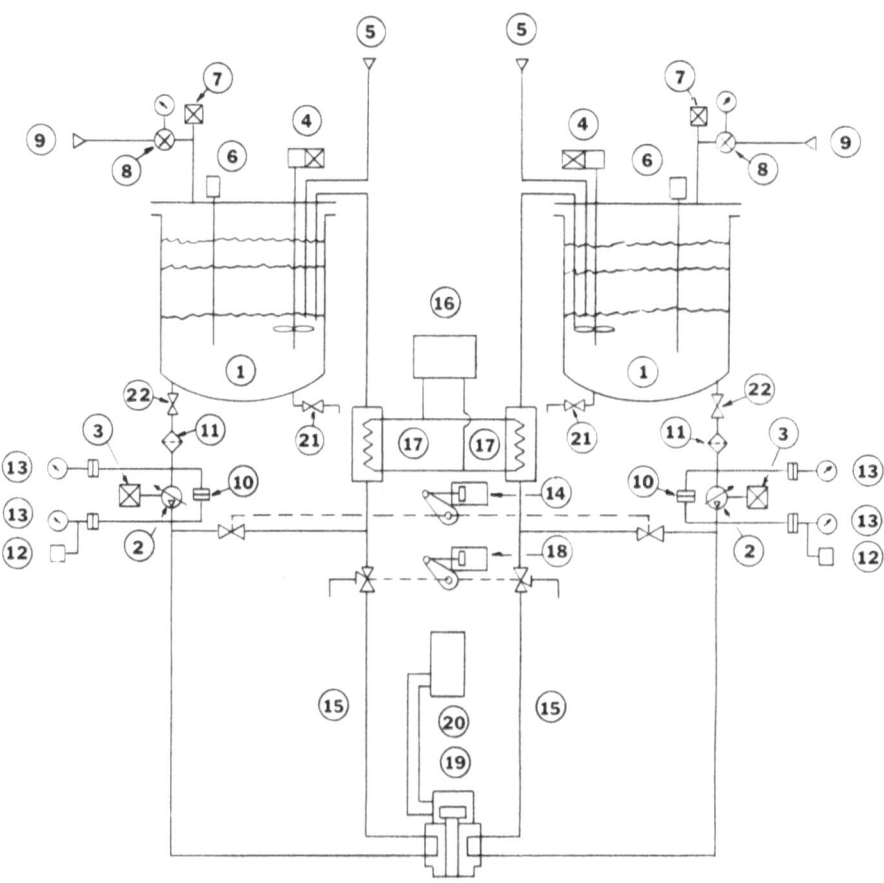

| REF. | QUAN. | DESCRIPTION | REF. | QUAN. | DESCRIPTION |
|---|---|---|---|---|---|
| 1 | 2 | TANK (2 GALLON STANDARD). | 12 | 2 | PRESSURE SWITCH (HIGH AND LOW SETTING). |
| 2 | 2 | HIGH PRESSURE METERING PUMP. | | | |
| 3 | 2 | 2 HP DC VARIABLE SPEED PUMP DRIVE. | 13 | 4 | PRESSURE GAUGE. |
| | | | 14 | 1 | LOW PRESSURE RECIRCULATE VALVE. |
| 4 | 2 | AIR MOTOR DRIVEN AGITATOR. | | | |
| 5 | 2 | REFILL INLET LINE. | 15 | 4 | HIGH PRESSURE HOSE LINES. |
| 6 | 2 | LEVEL CONTROL-INDICATION (3 LEVEL). | 16 | 1 | TEMPERATURE CONTROL UNIT (HOT & COLD SUPPLY) (OPTIONAL). |
| 7 | 2 | TANK RELIEF VALVE. | | | |
| 8 | 2 | DRY AIR PRESSURE REGULATOR & GAUGE. | 17 | 2 | SHELL & TUBE TYPE HEAT EXCHANGERS. |
| 9 | 2 | DRY AIR INLET LINE. | 18 | 1 | CALIBRATING DEVICE (OPTIONAL). |
| 10 | 2 | RUPTURE DISC (OVER PRESSURE PROTECTION). | 19 | 1 | HIGH PRESSURE MIXING HEAD. |
| 11 | 2 | HIGH PRESSURE INLINE FILTER. | 20 | 1 | HYD. POWER UNIT FOR HEAD. |
| | | | 21 | 2 | DRAIN VALVE. |
| | | | 22 | 2 | SHUT-OFF VALVE. |

Figure 11.  Schematic flow diagram and description of the "pta
martin sweets" high pressure metering and mixing
machine HP-20L.

tions in shot size; however, in regard to the Mini-RIM, a larger quantity of material as well as more extensive flushing is required for cleaning of the system due to the type of pumps employed and line system.

For the decision whether to employ a "mini-RIM" or a "HP-20L" type unit, the final use of the machine has to be observed.

# SIMULATION OF CAVITY FILLING AND CURING

# IN REACTION INJECTION MOLDING

L. T. Manzione

Bell Laboratories

Murray Hill, NJ 07974

## INTRODUCTION

Reaction injection molding[1] is polymerization and processing in a single cyclic operation. It consists of several distinct processes. Reservoirs of one, two, or more components are metered according to a pre-determined ratio and then conveyed under pressure to a mixing head. The fluids are either mixing activated in the head or thermally activated upon injection into the mold cavity. Considerable viscosity increase can occur during cavity filling, the extent depending on the fill time and resin reactivity. Most of the chemical conversion usually occurs during the mold curing step after the mold has been filled. The part is ejected after some fixed mold residence time which allows the part to attain dimensional stability. The cycle time includes the time for cavity filling, mold curing, and mold preparation and it is often desirable for it to be a minimum. Excessive cycle times due to long cure times may eliminate or limit the use of a molding resin in a particular application.

Material and process parameters determine the moldability of a resin in a particular application. Characteristic properties of a resin such as the gel time and the time to attain an ejectable rigidity are strongly influenced by temperature, both the temperature of the components and the mold temperature. Higher temperatures provide for more rapid reactions and shorter reaction times, but they can also increase the viscosity rise during cavity filling and the maximum exotherm temperature. Thermal runaway[1] is often encountered in reactive polymer processing and it is often desirable to minimize the cycle time.

This report is on a procedure to evaluate and then optimize different molding resins for different reactive processing applications and the use of this procedure to determine moldability. Cavity filling and curing, and in particular how they are influenced by resin and molding parameters, are investigated.


CAVITY FILLING

Simulation of reactive cavity filling is important in the determination of moldability because it allows prediction of·the viscosity and temperature rise.  It is not uncommon for the fluid to gel before the entire mold has been filled and this is known as premature gelation.[7]  Cavity filling simulation helps to avoid flow seizure; it can be used to optimize the process by allowing the most reactive resins to be molded, and it provides the initial conditions for the subsequent analysis of the cure in the mold cavity once filling has been completed.  A previous report on the flow of polymerizing fluids in reaction molding indicated that the process is sensitive to the fluid and kinetic parameters of the resin.[2]

Most reaction injection mold filling is under conditions of constant volumetric flow rate.  The alternate condition of a constant-pressure driven flow where the flow rate can vary with the viscosity of the fluid is similar to the present analysis and requires an additional convergence on the overall pressure drop.

The mold cavity is assumed to be of relatively simple geometric cross section.  Complex cavities can also be treated with the following technique or with other numerical methods such as finite elements.  The mold volume is divided into a large number of cells in accord with common finite difference methods of solution.  Both cylindrical and slab geometries result in axisymetric flows hence only two dimensional cells running the length of the cavity are needed.

Cavity filling was simulated using a modification of the simplified marker and cell (SMAC) method,[3,4,5] a finite difference method that is able to treat free surface fluid flows.  Marker and cell has previously been applied to cavity filling with thermoplastic melts[6] but the present effort is the first reported use of SMAC for reactive fluids.  The cells are characterized according to whether they are solid-fluid interfaces, free slip or no-slip boundaries, inflow or outflow surfaces.  In successive time increments, markers representing fluid elements enter and move through the cavity according to prescribed governing equations.  A fixed number of markers enters the cavity in the given time interval corresponding to the fixed volume flow rate.  A marker is associated with each filled cell and each marker carries information on the temperature and conversion of the fluid in that particular cell.

The temperature profile, for the usual case where a temperature gradient develops during filling, was computed from the following energy balance which describes the temperature change due to conduction, convection, and heat of reaction.

$$\rho C_p \left[ \frac{\partial T}{\partial t} + v_z \frac{\partial T}{\partial t} \right] = k_t \left[ \frac{\partial^2 T}{\partial x^2} + \frac{\partial^2 T}{\partial z^2} \right] + k_o \exp[-E_k/RT] C_A^n H_r \qquad (1)$$

The density, specific heat, and thermal conductivity were all assumed to remain constant. Empirical relations for the physical constants are preferable and they can be easily incorporated when they are available. An $n^{th}$ order kinetic expression was used to represent the rate of chemical reaction. The energy balance was then written as an explicit difference equation. The temperature field for transient conduction and chemical reaction (convective terms dropped) was in good agreement with previous reports using implicit methods,[7] various limiting conditions such as adiabatic reaction, and analytical solutions of conductive heat transfer.[8] The temperature of the mold wall was obtained by employing a similar finite difference solution, requiring a different mesh, in the mold wall. The interface temperature was determined with a convective film heat transfer coefficient found from simple heated tube flow experiments and empirical correlations[9] where applicable.

The concentration field was obtained by writing the following mass balance on a single component. The stoichiometric ratio should also be included if a stoichiometric imbalance is present.

$$\frac{\partial C_A}{\partial t} + v_z \frac{\partial C_A}{\partial z} = D_{AB}(X,T) \left[ \frac{\partial^2 C_A}{\partial x^2} + \frac{\partial^2 C_A}{\partial z^2} \right] - k_o \exp[-E_k/RT] C_A^n \qquad (2)$$

The mass balance was also written as an explicit finite difference equation and solved simultaneously with the energy balance for the concentration and temperature field.

The viscosity of the fluid in each cell can then be calculated from the chemical conversion and the temperature using one of the following empirical relations:[10,11]

$$\mu = \mu_\infty \left[ \frac{k_1 X_{gel}}{X_{gel} - X} \right]^{k_2 X} \exp(E\mu/RT) \qquad (3)$$

$$\mu = \mu_\infty \exp[E_\mu/RT] \exp[X] \qquad (4)$$

The relation shown in Equation (3) (after Castro and Macosko[10]) was used in the current work. These constitutive relations for thermoset or polymerizing materials are relatively simple in that they do not include any non-Newtonian or elastic behavior. Although this assumption is debatable for some materials at later stages of cure or near gelation, it is usually valid for most low viscosity resins at low conversion. More complex constitutive relations for reacting fluids could probably be incorporated in the model in a manner similar to an earlier use of the marker and cell method in the simulation of mold filling thermoplastic fluids.[6]

The velocity field can be computed once the viscosity field is known. The following momentum balance for the axial component of the velocity behind the front was written:

$$\frac{\partial^2 v_z}{\partial x^2} + \frac{1}{\mu(x)} \frac{d\mu}{dx} \frac{dv_z}{dx} = \frac{1}{\mu(x)} \frac{dp}{dz} \tag{5}$$

The temporal term was dropped since this expression is applied at each time interval and in that way responds to the time-dependent viscosity field. The transverse velocity ($v_x$) was obtained through continuity assuming a lubrication flow since the aspect ratio of the tube is large.

$v_z$ and $v_x$ velocity components were again obtained through a finite difference solution and the components combined into a representative velocity vector. The markers are then moved into new cells in response to their velocity vectors and the duration of the time interval. Markers near the center of the cavity, for example, usually have larger axial velocities than those near the wall and they move the greater distance. Cells near the gate of the mold, vacated by markers that have moved further into the cavity, are filled by markers moving in through the gate.

Markers near or at the front would project out into the empty cavity. These markers, however, are required to follow the streamlines that define a fountain flow, a transient fluid flow pattern found in the filling of thin cavities with viscous fluids. The streamlines were arrived at from previous reports on fountain flow,[12,13,14] and two component injection molding.[15] The streamlines are shown in Figure 1. This treatment of fountain flow, similar to the heuristic treatment of Domine and Gogos,[2] and identical to that of Lord and Williams,[14] allows the incorporation of the effects of fountain flow on the residence time distribution yet avoids the mathematical difficulties associated with the moving contact line. Variation of the stream pattern used did not cause any noticeable difference in the resulting temperature and conversion profiles found after cavity filling. This indicates that an exact solution of the fountain flow is probably not required for this type of

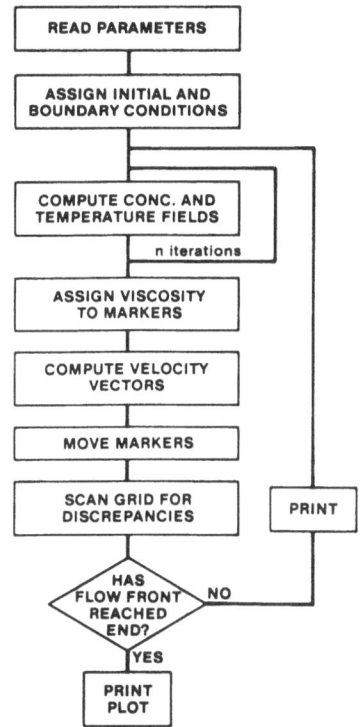

Figure 1.    Streamlines in the front region characteristic of
fountain flow.

moldability analysis as long as it is considered.

Once all the markers have been moved, a scan is made of all the
cells behind the new front to find any cells that do not have markers or cells that have two or more markers.  A new marker is created
in an empty cell by averaging the information of adjacent markers.
Multiple markers in a single cell are likewise averaged to produce
a single representative marker.  A new cycle is started when this
scan is complete.  The time increment counter is increased and a
new temperature, concentration, viscosity and velocity field found
from the coupled balance equations.  The markers are moved, new markers enter the cavity, and the entire grid scanned for discrepancies.
A flow chart of the cavity filling program is presented in Figure 2.

The velocity profile at any cross section of the cavity must
provide the given volumetric flow rate.  The pressure drop serves as
the normalizing parameter to satisfy this important condition of the
incompressible flow.  The cycle is repeated as long as the results
are of interest or until the flow front has reached the end of the

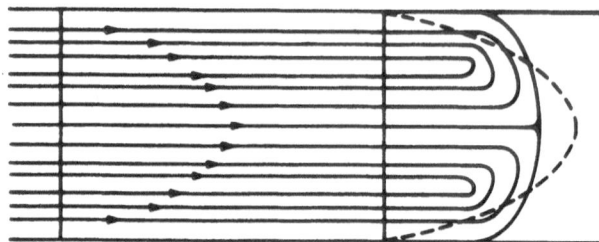

Figure 2.   Flowchart for the cavity filling program.

mold cavity.   Temperature, conversion, viscosity and velocity profiles within the cavity at each time increment are obtained in this
manner.

MOLD CURING

Temperature and conversion profiles in the mold cavity after
filling is complete are obtained by again solving the coupled energy
and material balance numerically with the specific boundary conditions
that are to be imposed and the appropriate parameters of the molding
resin.   The material is stationary, no convective terms are used, and
no markers are moved.   The heat capacity, thermal conductivity, and
the density are all assumed to remain constant as a first approximation.   Empirical relations for physical properties are preferable
and are easily accommodated if available.   In general, the same procedure applied in the filling simulation is continued until the material in the cavity is at full conversion or has reached an ejectable
conversion.

Stonecypher et al[16] have solved similar dimensionless equations
for the case of an isothermal wall and other considerations in the
casting of large parts.   Macosko and Broyer[7] have also solved for
conversion profiles during curing with various isothermal and adiabatic boundary conditions for typical reaction molding resins.   The
current effort considers the general reaction molding problem since
a numerical solution was obtained for the temperature and conversion
profiles that incorporate the curing that occurs during cavity filling.   The mold wall was treated in the same manner as in the filling
analysis; that is, the separate finite difference solution for the
temperature profile in the mold wall that couples to the solution
inside the cavity was continued to provide a realistic boundary condition.   The results obtained when uniform initial conditions were
assumed were in good agreement with previous reports[7] and with the
various limiting conditions such as adiabatic molding.   The initial
conditions of conversion and temperature are established from the
flow of the reactive fluid into the mold cavity as previously derived.
In the absence of such coupling, there is no axial variation in the

profiles down the length of the cavity and significant differences
are found in the origin and nature of the thermal runaway that is
observed.


RESULTS

    The reactive fluid used here for demonstration purposes is sim-
ilar to a conventional RIM urethane.  A graphic depicting the viscosity as a function of conversion and temperature is shown in Figure 3.  Subsequent alteration of some properties to explore limits
and criteria for moldability will be clearly indicated.  A cavity
geometry similar to a common plaque mold was used.  The gate area
is the entire left-half side of the mold and the fluid front moves
from left to right.  The cavity is assumed to be sufficiently wide
such that end effects are not considered.  After the flow has stopped,
the gate is treated as a solid wall and the boundary conditions there
and at the far wall were derived from the same finite difference solution used for the top and bottom faces.  The cavity was nominally
12 cm long and 0.6 cm thick.  The average fluid velocity was 1.00
cm/sec which provided a fill time of 12 seconds.  This relatively
low velocity was used to promote sufficient conversion in the short
cavity that was required by computer hardware considerations.

    The first conditions considered are typical of reaction mold-
ing; a component temperature of 30°C and a wall temperature of 70°C.

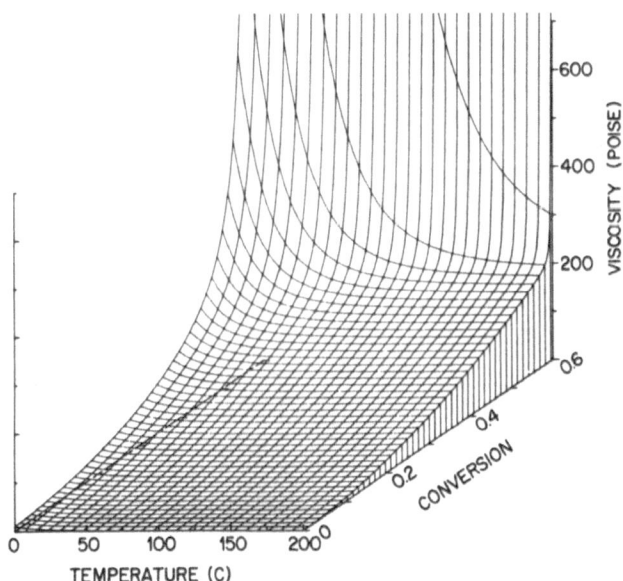

Figure 3.  Viscosity as a function of conversion and temperature.

Conversion and viscosity profiles in the mold cavity at the time
the flow front reaches the end of the mold are presented in Figures
4 and 5. The axial velocity profiles at this same instant are presented in Figure 6. Unfortunately, this final velocity profile cannot show the variety of changing profiles occurring throughout the
stages of cavity filling. Plug-like laminar profiles characterize
the early stages of filling where fresh, low conversion fluid first
contacts the heated wall. The viscosity of this fluid drops and the
resulting wall region velocity is greater than that of a parabolic
profile which is the baseline condition with uniform viscosity.
Note that the oldest, most viscous fluid is located at the wall and
nearer to the entrance of the cavity. These results are in quali-

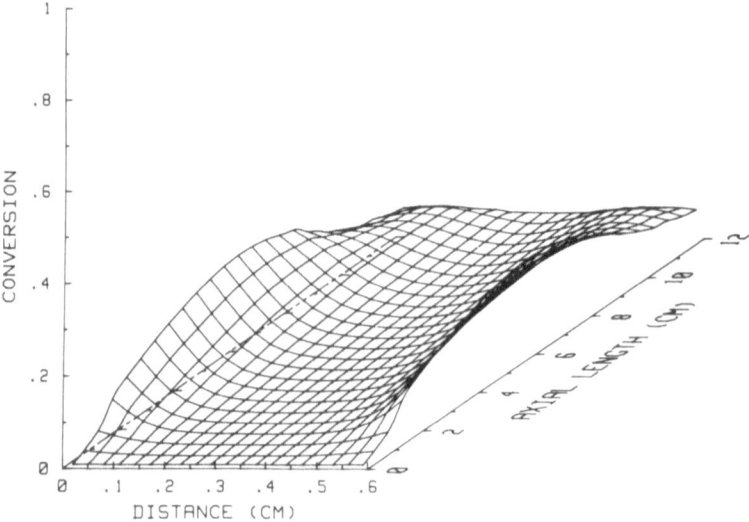

Figure 4.   Conversion profiles after cavity filling ($T_o$ = 30°,~
            $T_{w_o}$ = 70°C).

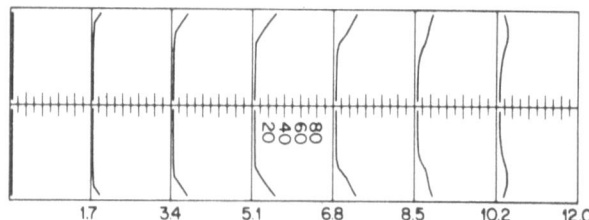

Figure 5.   Viscosity profiles after cavity filling ($T_o$ = 30°,
            $T_{w_o}$ = 70°).

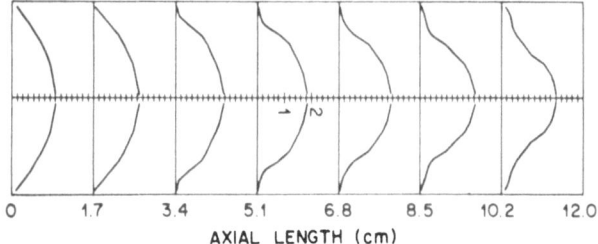

Figure 6.   Velocity profiles as front reaches the end of the
            cavity ($T_O$ = 30°, $T_{W_O}$ = 70°).

tative agreement with the tracer studies of cavity filling by
Schmidt[17] where iso-age tracer material was found in reverse order
to how it entered the cavity (material in first was found near the
gate) and drawn into long V shapes.  The center of the channel contains the fresher, low viscosity fluid that had recently entered the
cavity; this is clearly evident in the 3-D graphics shown in Figure
4.  Notice how the high conversion region encroaches on the center
and how the conversion of this central fluid is also slowly increasing down the length of the cavity (see Figure 5).

     There are several important implications of this result.  The
most obvious is that the higher conversion fluid is not located at
the front.  A more important result is that even though a region
of high conversion will usually develop at the wall somewhere down
the length of the cavity, flow seizure will not occur there due to
the relatively fresh, low temperature fluid in the center of the
channel at that axial position.  The flow will be halted only when
gelation occurs at the front since this is the oldest central fluid.
Another observation, evident in the velocity profiles down the length
of the cavity, is that although the flow will not seize at the high
conversion area, a growing gel layer at the wall may cause the flow
to nozzle and jet into the cavity.

     Temperature and conversion profiles at regularly spaced posi-
tions in the cavity at increasing times after filling is complete
are presented in Figures 7 and 8.  The time increment between plotted lines is 1 second and lines after the maximum is reached at that
axial position are omitted for clarity.  The initial condition is
the first or lowest conversion or temperature line shown.  It is the
condition found at the instant the mold is filled and is the two
dimensional analog of Figure 4.  The time to reach full cure or the
mold residence time is evident as the time or number of lines required for the conversion to reach an ejectable level and in this
case is an additional 19 seconds after the 12 second fill time.

     Full conversion is reached at different times at different

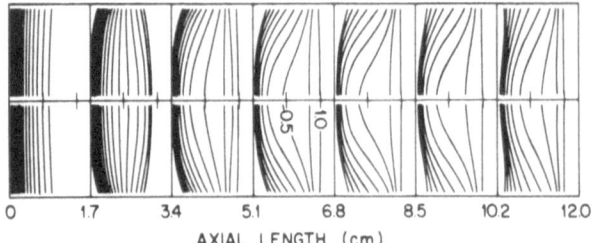

Figure 7.   Conversion profiles in the cavity during mold curing.
Time increment between plotted lines is 1 second.

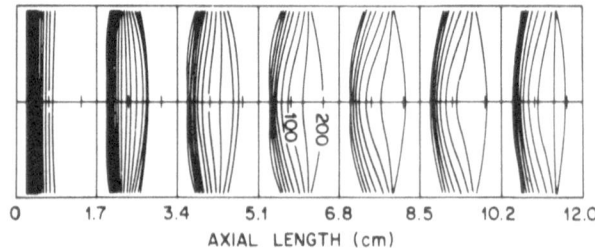

Figure 8.   Temperature profiles in the cavity during mold curing.
Time increment between plotted lines is 1 second.

positions in the mold cavity.  This reflects the differences in
cure resulting from the flow of the reactive fluid.  As in the present case, the gate region is usually the last to reach full conversion because the freshest fluid is found there at the termination
of filling.  Initially, higher temperatures are found near the mold
wall since the fluid there is older, has been heated, and the reaction activated.  This increased wall temperature often persists during cure.  Full conversion may be reached first at the center of
the mold, however, due to the near adiabatic conditions present
there.  These adiabatic conditions result in a thermal runaway that
is clearly evident in the growing temperature and conversion differences between the isochrones of Figures 7 and 8.  Thermal runaway
is the condition where the product of the temperature and concentration terms in the reaction rate is a maximum.  It is a condition
often found with exothermic, highly activated reactions that is
similar to an explosion.  The present results indicate that thermal
runaway is important in reaction injection molding because it provides for the rapid reactions and short cycle times that allow the
process to compete with other polymer and metal processing techniques.  Thermal runaway is fostered by adiabatic conditions and
is, therefore, suppressed by enhanced heat loss; thinner cavities,
more conductive mold materials, or lower mold wall temperatures.

Thermal runaway is not a characteristic of the molding resin or of the process parameters but of a combination of the two that can only be predicted by a detailed analysis that considers their interaction[1] or a complete simulation of the reaction path as in the present case. Thermal runaway in one region of the mold cavity does not insure that it will spread over the entire cavity. The low thermal conductivity of polymers tends to isolate the exotherm, particularly if the cavity is wide and long, and the gate area often lags in reaching full cure. This may not be encountered in practice, however, since a short circuit is afforded by the relatively rapid conduction through the mold itself. The complete solution of the temperature profile in the mold wall correctly simulates this phenomenon and a moderate runaway is usually found in the gate region. In cases of very rapid reactions, however, the mold may act as a thermal mass that lags and inhibits the rapid temperature rise within the cavity.

An important criteria in reaction molding is the amount, if any, of external heat required for satisfactory molding in a reasonable cycle time. This heat can be supplied directly by heating the components prior to mixing or by heating the mold. The common temperature of the components is denoted $T_o$ and the wall boundary condition is $T_{w_o}$. These can be varied independently and the resulting temperature and conversion profiles at the time the cavity is filled reveal the extent of cure during filling.

The extent of cure is far more sensitive to the component temperature than to the wall temperature but this lack of sensitivity to $T_w$ decreases as the width of the cavity decreases. Temperature profiles after filling resulting from a wall temperature of 110°C (compared to 70°C presented in Figure 5, all else is unchanged) are shown in Figure 9. There is a considerable increase in the conversion that has occurred during filling, yet thermal runaway and flow seizure has not occurred. The overall cycle time for the conditions would be shorter because greater conversion occurred during the fixed filling time. In this way, the simulation can be used to predict the greatest wall temperature that allows the cavity to fill, thereby minimizing the overall cycle time.

The curing analysis for these conditions is shown in Figure 10. The reaction at the far end of the cavity is rapid and it goes to completion in 6-8 seconds. The cure near the gate lags and it requires 16 seconds to reach even an ejectable conversion of 60%. The overall time of reaction is too short to take advantage of the heating of the mold by the reaction exotherm to activate the gate region, so the additional conversion at the far end of the cavity has contributed little to reducing the overall cycle time. The mold surface will cool as low temperature fluid passes over it during filling and there is a lag before it recovers and is heated by the exotherm.

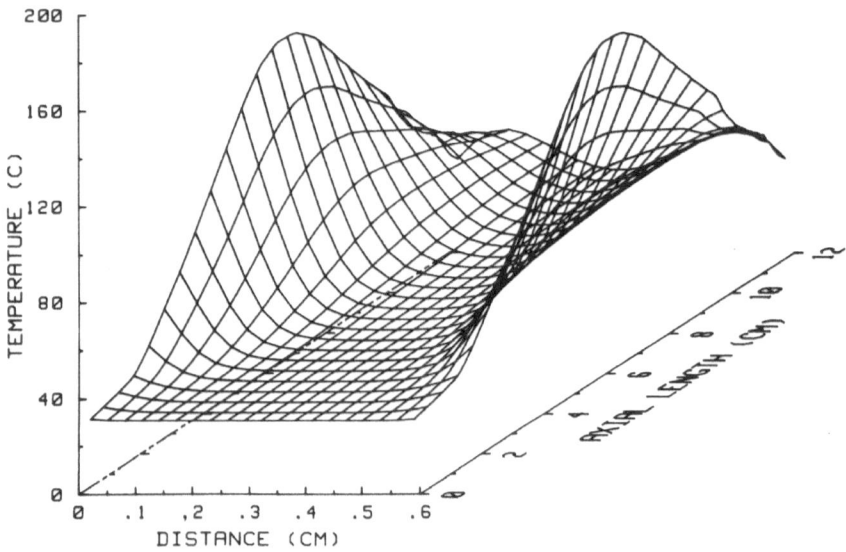

Figure 9.    Temperature profiles after cavity filling ($T_0$ = 30°C,
$T_{w_0}$ = 110°C).

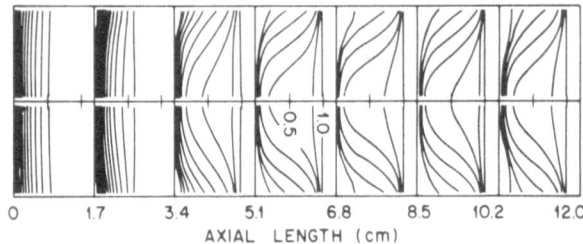

Figure 10.    Conversion profiles in the cavity during mold curing.
($T_0$ = 30°C, $T_{w_0}$ = 110°C).    Time increment between
plotted lines is 1 second.


Temperature profiles resulting from cavity filling when the
initial temperature of the components has been increased from 30°C
to 40°C ($T_{w_0}$ = 70°) are shown in Figure 11.    The temperature tracks
closely to the conversion in most cases and occasionally local "hot
spots" are found in both.    Note the large extent of conversion in
Figure 11 – considerably more than in Figure 4 ($T_{w_0}$ = 70°, $T_0$ = 30)
and comparable to Figure 9 ($T_{w_0}$ = 110°, $T_0$ = 30).    In particular,
there is a considerable increase in the centerline conversion and
temperature at the end of the cavity.    The centerline conversion
and temperature are far more important than the wall conditions
since they precipitate thermal runaway.    Subsequent analysis of cure
in the mold cavity after filling is completed (see Figure 12) shows

that the central fluid gels in just an additional 2-4 seconds indicating that the fluid would flow only an additional 4 cm under these
conditions.  Once again, the reaction at the far end of the cavity
is very rapid and it reaches full conversion first.  The cure near
the gate region still lags but it is faster than the results for
($T_W$ = 70, $T_O$ = 30) or ($T_{W_O}$ = 110, $T_O$ = 30).  An additional 13 seconds is needed before the part can be ejected.  The clear implica-
tion here is that heating of the components is much more effective
in promoting conversion during filling and minimizing the cycle
time than heating of the mold wall.  This is reasonable in light of
the high activation energy and low thermal conductivity of most
reaction molding resins.  In this way, the simulation can be used
to optimize the component temperature to minimize the overall cycle
time.  In general, though, greater conversion during filling does
not guarantee a shorter cycle time because the conversion is usually non-uniform and therefore localized areas of fluid may remain
unactivated and lag behind the rest of the cavity.

CONCLUSIONS

A procedure was developed to evaluate and then optimize the
reaction injection molding of thermoset resins in various applications.  The method consists of several mathematical simulations that
are easily adapted to handle a number of simple geometries.  Cavity
filling can be simulated with the marker and cell method to provide
results that agree qualitatively with previous experimental and

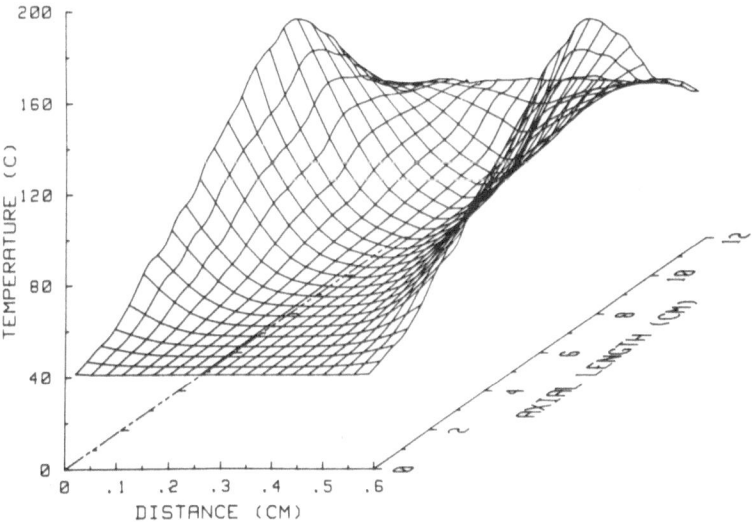

Figure 11.   Temperature profiles after cavity filling ($T_O$ = 40°C,
             $T_{W_O}$ = 70°C).

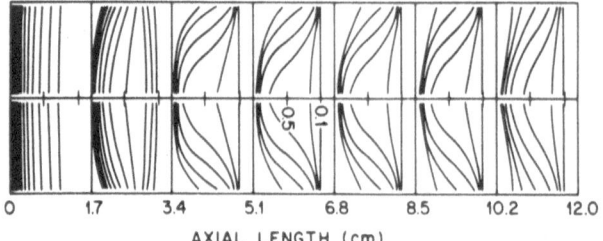

Figure 12.  Conversion profiles in the cavity during mold curing.
($T_O$ = 40°C, $T_{w_O}$ = 70°C).  Time increment between
plotted lines is 1 second.

tracer studies.  The results indicate the complexity and non-uni-
formity of the temperature and conversion profiles that occur during and after cavity filling with reactive fluids.  It appears flow
seizure results when the central fluid at the front approaches gelation yet the ability to fill is fairly insensitive to gelation at
the wall.  In general, it is advantageous to promote more conversion during cavity filling since it can reduce the overall mold
residence time.

The cure in the mold cavity after filling is complete can then
be modelled using conventional finite difference methods.  These
results elucidate the phenomenon of thermal runaway and its importance in reaction injection molding.  The overall analysis is unique
because it retains the coupling between mold filling and mold curing that has often been reported but not accounted for in predict-
ing cycle times and maximum exotherms.  This coupled analysis can
then be used to predict the occurrence of flow seizure during cavity filling, the maximum exotherm temperature, and the demold time.


## Acknowledgements

The author is grateful to Julian S. Osinski for developing the
3-D graphics presented in this work as well as for helpful discussion during its development.


## Nomenclature

$C_A$   – reactant concentration

$C_{Ao}$  – initial reactant concentration

$C_p$   – specific heat

$D_{AB}$ — diffusivity

$K_k$ — activation energy for chemical reaction

$E_\mu$ — activation energy for fluid flow

$H_r$ — heat reaction

$k_i$ — empirical constants for viscosity relation; $i = 1,2$

$k$ — reaction rate constant

$k_o$ — frequency factor for chemical reaction

$k_t$ — thermal conductivity

$n$ — order of chemical reaction

$p$ — pressure

$R$ — gas constant

$t$ — time

$T$ — temperature

$\nu$ — velocity

$x$ — chemical conversion

$x_{gel}$ — conversion at gelation

$W$ — half-cavity thickness

$\rho$ — density

$\mu$ — viscosity

$\mu_\infty$ — viscosity at "infinite" temperature

## References

1. J. A. Bisenberger and C. G. Gogos, Polym. Eng. Sci. 20 (13), 838 (1980).
2. J. D. Domine and C. G. Gogos, Polym. Eng. Sci. 20 (13), 847 (1980).
3. F. H. Harlow and J. E. Welch, Physics of Fluids, 8 (12), 2182 (1965).

4.  J. E. Welch, F. H. Harlow, J. P. Shannon and B. J. Daly, U. S. Govt. Report (NTIS LA-3425) (1966).
5.  A. A. Amsden and F. H. Harlow, U. S. Govt. Report (NTIS LA-4370) (1970).
6.  C. F. Huang, Ph.D. Thesis, Stevens Institute of Technology, Dept. of Chemical Engineering (1978).
7.  E. Broyer and C. W. Macosko, A.I.Ch.E. J. 22 (2), 268 (1976).
8.  B. Carnaham, H. A. Luther and J. A. Wilkes, Applied Numerical Methods, John Wiley & Sons, Inc., New York, 1969, p. 432.
9.  A. H. P. Skelland, Non-Newtonian Flow and Heat Transfer, Wiley (1967).
10. J. M. Castro and C. W. Macosko, S.P.E. Technical papers, ANTEC, New York, NY, May, 1980 (p. 434).
11. K. M. Hollands and I. L. Kainin, Adv. in Chem. Ser. No. 92, 60, American Chemical Society, Washington, DC (1970).
12. Z. Tadmor, J. Appl. Polym. Sci., 18, 1753 (1974).
13. E. Broyer, C. Gutfinger and Z. Tadmor, Trans. Soc. Rheol., 19 (3), 432 (1975).
14. H. A. Lord and G. Williams, Polym. Eng. Sci., 15 (8), 569 (1975).
15. R. C. Donovan, K. S. Rabe, W. K. Mammel and H. A. Lord, Polym. Eng. Sci., 15 (11), 774 (1975).
16. T. E. Stonecypher, E. L. Mastin Allen and D. A. Willougby, Chem. Eng. Prog. Symp. Series 62, 7 (1966).
17. L. R. Schmidt, Polym. Eng. Sci., 14 (11), 797 (1974).

# COMPUTER ANALYSIS OF RIM MOLDABILITY FROM THE

# DENSITY DISTRIBUTION IN MOLDED PRODUCTS

Kazuo Okuda

Research & Technical Service Division
Kasei Upjohn Company
Yokohama-City, Kanagawa, Japan

## INTRODUCTION

Under the worldwide campaign to "save resources and energy," automobiles have been improved by reducing weight and increasing fuel efficiency. In Japan the use of urethane bumpers which meet these objectives has been steadily increasing. However, the manufacture of polypropylene bumpers as a substitute is also increasing, due to cheaper starting materials. The RIM urethane technology is very well suited to the production of parts that are large in size and complicated in shape. The problems with moldability arise when parts are made thinner, larger and when cycle time is reduced. These problems connected with moldability can be solved to some extent by optimization of the molding conditions, by better mold design and by selection of a better urethane system. However, the development of a new urethane system requires many experimental trials to optimize it because of the many variables involved, such as catalysts, isocyanates, polyols, blowing agents, temperature of starting materials, etc. At the present time there are only a few papers in literature discussing problems associated with the moldability of RIM systems. In this paper, computer techniques suitable for the optimization of physical properties and moldability of RIM systems will be discussed.

## PROBLEMS IN URETHANE BUMPER PRODUCTION

One of the major problems in RIM bumper production are the molding problems that are associated with the product thickness. Figure 1 shows the density contours of the urethane bumper drawn by the computer. As can be seen from this figure, the contour lines of

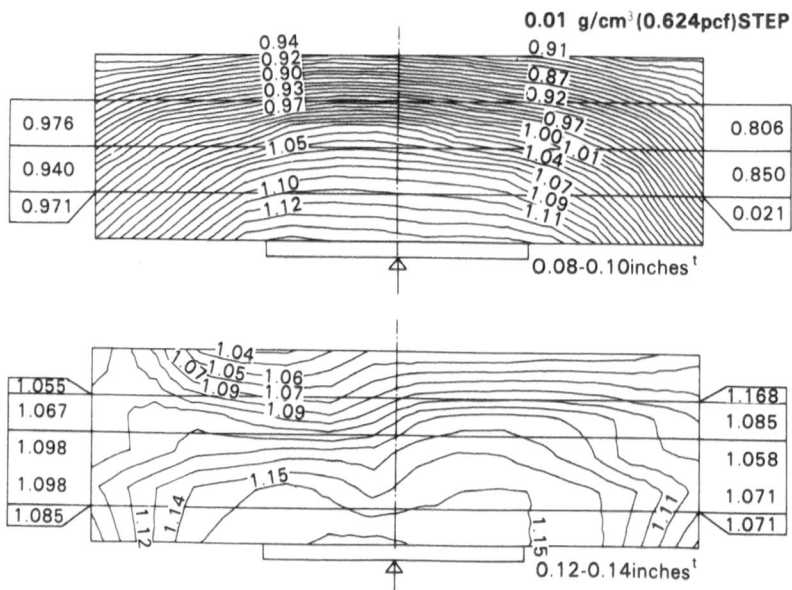

Figure 1.  Density distribution of bumper products (effect on the
bumper thickness).

$0.01$ g/cm$^3$ steps are very closely arranged together in the case of
a thinner type of bumper, especially at the corners where the density becomes so low that it is difficult to make a good product.

Figure 2 shows the typical molding defects in bumper production:

1.  Sink Marks.  The poor flowability causing high density near
the gate generates uneven surfaces in striped or round shapes which
are called sink marks.  The sink marks become round when there are
cases like chip spots of the releasing agent.  The reason for the
occurrence of sink marks is unclear, but they can easily be caused
by high density and closely arranged high density distribution.  The
local difference of the conversion of liquid injection seems to
cause this phenomena.

2.  Loose Skins or Flow Marks.  A less reactive urethane system
brings about loose skins or peeling of the surface near the gate or
at the corner, causing stains on the mold.  The flow marks are the
radial loose skins near the film gate.  This can easily be caused
when the shape of the film gate is not adequate.  The loose skins
are often seen when the reactivity is poor.  These phenomena are
generally caused more often in ethylene glycol-based systems than
in 1,4-butanediol-based systems.

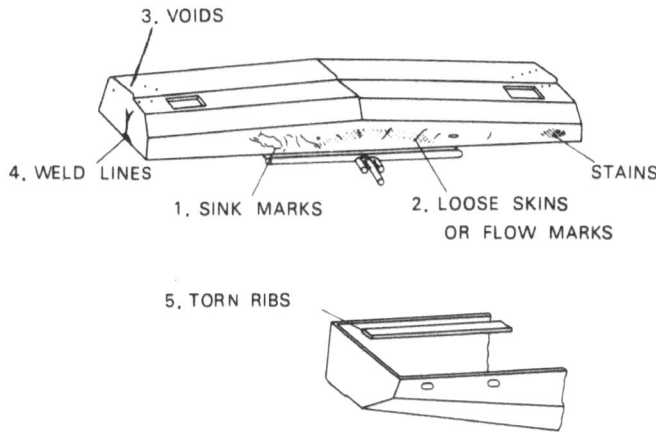

Figure 2.  Molding defects in the production of bumpers.


    3.  Voids or Bubbles.  In the case of less flowable systems,
voids behind the obstacles and small bubbles caused at the film gate
by sucking air sometimes appear on the surface of the product.  These
faults result in outside expansions on the surface when the product
is heated for painting.

    4.  Weld Lines.  Many small lined bubbles occur at the end of
the flow when the gel time starts before the liquid injection has
filled up the mold.  The crack of the product occurs when this
phenomena is intense.  This also occurs when enough air does not
escape from the mold.  This fault deteriorates physical properties
near the end of the product.

    5.  Tearing the Rib or Hole-Side.  The lower reactivity of a
system and low densities of RIM products near the ribs due to less
flowability decrease the green strength of the product and the ribs
or hole-side are torn by the frictional force between the mold and
the product during demolding.

    While the above-mentioned problems are the major ones, there
are some additional molding difficulties related to urethane systems.
Some of these are expansion at the back side of the rib, poor mixing
near the injection gate and deformation after demolding.

    From our experimental point of view the problems of molding
depend on two factors, that is, the flowability and the reactivity.
As shown in Table I, a higher productivity could be achieved by the
improvement of these two factors.  Good reactivity of RIM systems
would reduce the demolding time and simplify the cleaning operation.
Good flowability would decrease the amount of substandard products.
However, the good reactivity results in poorer flowability by short-

Table I.  Molding Problems in RIM Bumper Production

```
REACTIVITY OF SYSTEM
        GREEN STRENGTH·····················Shortening the demold time
        LOOSE SKINS                    ⎤  Shortenig the cleaning
        FLOW MARKS                     ⎮  time for the mold
        STAINS ON THE PRODUCT          ⎦
FLOWABILITY
        SINK MARKS                     ⎤  Decreasing the substandard
        VOIDS                          ⎮  products
        WELD LINES                     ⎦
```

ening the gel time.  Therefore, the flowability and the reactivity
seem to be inconsistent with each other.  But the viscosity of the
reactants is reduced by the exothermic reaction in the course of
polyaddition reaction as Broyer clarified.  Therefore, various
systems indicate many flow patterns and characteristics.  Up to now,
however, there are few reports available covering practical moldability.  The purpose of this paper is to describe a new computer
method suitable for the screening of RIM urethane bumper systems.


SCREENING OF RIM SYSTEMS FOR OBJECTIVE PHYSICAL PROPERTIES

     The first method to be discussed is that of screening of RIM
systems which fit the objective physical properties by means of a
computer.  Usually many physical properties, such as high temperature
properties, low temperature properties and strength, must be taken
into account in RIM systems.  Polyols, chain extenders and isocyanates formulated in adequate quantity make the best system to give
objective physical properties.  However, many experimental trials
are needed to screen the systems.  Applying the same technique used
by Carleton, mathematical models of the regression analysis between
systems and physical properties were created.  In the case of ethylene glycol-based systems, we usually blend some polyols to improve
the moldability.


REGRESSION ANALYSIS

     Three kinds of polyols, such as polyol A (mw. 6500 triol),
polyol B (mw. 2400 diol) and polyol C (mw. 3000 triol) were used
as the experimental examples.  The three polyols were mixed at
various ratios and ethylene glycol was used as chain extender.  Then
catalysts, blowing agents and carbon pastes were added.  The samples
were made with a RIM machine for various systems and physical properties were respectively measured.  Then we did regression analysis
between the systems and the physical properties using the BMD-02D
program made by UCLA.

The three kinds of polyols were taken into account by use of Equation 1:

$$\overline{POH} = POH_a \cdot W_a + POH_b \cdot W_b + POH_c \cdot W_c$$

$$\overline{MW} = MW_a \cdot W_a = MW_b \cdot W_b + MW_c \cdot W_c$$

$$\overline{OH} = OH_a \cdot W_a + OH_b \cdot W_b + OH_c \cdot W_c$$

(1)

From our experimental results it became clear that the number of functional groups and the ratio of ethylene oxide or the ratio of primary OH in the end group of the polyol intensively affected the physical properties.  Thus these two factors were used here as independent variables for the regression analysis.  We consequently used seven variables and eight interacting variables for stepwise regression analysis as shown in Table II.

EXPERIMENTAL RESULTS OF REGRESSION ANALYSIS

The 43 data were used for the regression analysis and the regression coefficients were obtained respectively for the individual physical properties as shown in Table III.  The regression equation for the flexural modulus is shown below:

Flexural modulus : FM $(kg/cm^2)$ =

$$-26.9 \ POH - 0.195 \ MW + 356 \ OH - 5.20 \ OH^2 + 3050 \ DEN + 7250 \ W + 9150 \ I - 101 \ RAT \qquad (2)$$

The errors between calculated values using regression equation and measured values were very small.  Table IV shows the correlation results for the brittleness temperatures.

Table II.   Independent Variables Used in Regression Analysis

1  $\overline{POH}$ ; average primary—OH ratio of polyol

2  $\overline{MW}$ ; average molecular weight of polyol

3  $\overline{OH}$ ; average OH value of polyol

4  W   ; ethylene glycol (pbw)

5  I   ; index

6  RAT ; mixing ratio (polyol/Isocyanate)

7  DEN ; density of elastomer

8  $I^2$, $I^3$, $OH^2$, $DEN^2$, $W^2$, I·DEN

   I·DEN·W,  W·DEN

Table III.   Regression Coefficients Obtained From 43 Systems

| | POH | MW | OH | OH$^2$ | DEN | W | I | RAT |
|---|---|---|---|---|---|---|---|---|
| SHORE D    - | -0.0819 | -0.00397 | 2.37 | -0.0374 | 40.7 | -13.7 | 20.4 | -0.156 |
| FLEXURAL MODULUS kg/cm$^2$ | 26.9 | -0.195 | 356 | -5.20 | 3050 | -7250 | 9150 | -101 |
| 50% MODULUS kg/cm$^2$ | -1.48 | -0.00122 | 163 | -0.230 | 356 | -979 | 244 | -4.30 |
| ULTIMATE STRENGTH kg/cm$^2$ | -1.16 | 0.00839 | 15.8 | 0.385 | 515 | -329 | 124 | -1.86 |
| ELONGA- TION % | 1.21 | -0.0231 | 15.6 | -0.286 | 343 | -889 | 108 | 0.214 |
| TEAR STRENGTH kg/cm | 1.04 | -0.0285 | 1.31 | -0.0452 | 111 | 5.18 | 32.2 | -0.144 |
| BRITTLENESS TEMP. °C | 0.190 | -0.00435 | -0.221 | -0.00871 | -34.5 | 76.5 | 28.6 | -0.334 |
| HEAT SAG 100°Cx1 hr mm | - | - | 0.109 | - | -51.3 | 152 | -22.2 | 0.515 |

Table IV.  Measured Brittleness Temperatures and Calculated
Values Using Regression Equations

| SYSTEM NUMBER | MEASURED VALUES | CALCULATED VALUES | RESIDUAL |
|---|---|---|---|
| 1 | 39.0 | 39.2 | -0.2 |
| 2 | 50.2 | 48.3 | 1.9 |
| 3 | 36.6 | 35.6 | 1.0 |
| 4 | 37.4 | 36.3 | 1.1 |
| 5 | 38.2 | 40.9 | -2.7 |
| 6 | 35.8 | 35.8 | -0.0 |
| 7 | 27.8 | 31.4 | -3.6 |
| 8 | 34.2 | 36.0 | -1.8 |
| 9 | 39.4 | 39.4 | -0.0 |
| 10 | 41.2 | 41.3 | -0.1 |
| 11 | 42.0 | 43.4 | -1.4 |
| 12 | 42.8 | 42.6 | 0.2 |
| 13 | 45.0 | 43.2 | 1.8 |
| 14 | 42.6 | 43.2 | -0.6 |
| 15 | 43.9 | 43.3 | 0.6 |
| 16 | 37.4 | 37.0 | 0.4 |
| 17 | 32.6 | 33.4 | -0.8 |
| 18 | 31.4 | 33.9 | -2.6 |
| 19 | 43.8 | 48.7 | -4.9 |
| 20 | 42.6 | 44.0 | -1.4 |
| 21 | 31.8 | 30.3 | 1.5 |
| 22 | 33.0 | 31.1 | 1.9 |
| 23 | 29.8 | 30.3 | -4.9 |
| 24 | 33.4 | 31.9 | 1.5 |
| 25 | 32.2 | 31.1 | 1.0 |
| 26 | 46.6 | 43.9 | 2.7 |
| 27 | 47.8 | 47.4 | 0.4 |
| 28 | 43.0 | 44.3 | -1.4 |
| 29 | 35.4 | 35.1 | 0.3 |
| 30 | 47.0 | 48.7 | -1.7 |
| 31 | 44.2 | 47.1 | -2.9 |
| 32 | 42.6 | 44.0 | -1.5 |
| 33 | 41.8 | 41.6 | 0.2 |
| 34 | 48.2 | 47.1 | 1.0 |
| 35 | 45.0 | 44.7 | 0.3 |
| 36 | 43.8 | 42.6 | 1.2 |
| 37 | 45.8 | 45.5 | 0.3 |
| 38 | 42.6 | 43.3 | -0.7 |
| 39 | 48.8 | 45.6 | 3.2 |
| 40 | 47.8 | 45.1 | 2.7 |
| 41 | 45.0 | 44.5 | 0.5 |
| 42 | 46.8 | 45.3 | 1.5 |
| 43 | 47.6 | 45.9 | 1.7 |

The effect of the composition of polyols, index, level of ethylene glycol on physical properties can easily be learned by using the
regression equation within the limits of the data.  Figure 3 shows
these relationships for a certain system.  It could be easily understood that this system shows the flexural modulus increase by about

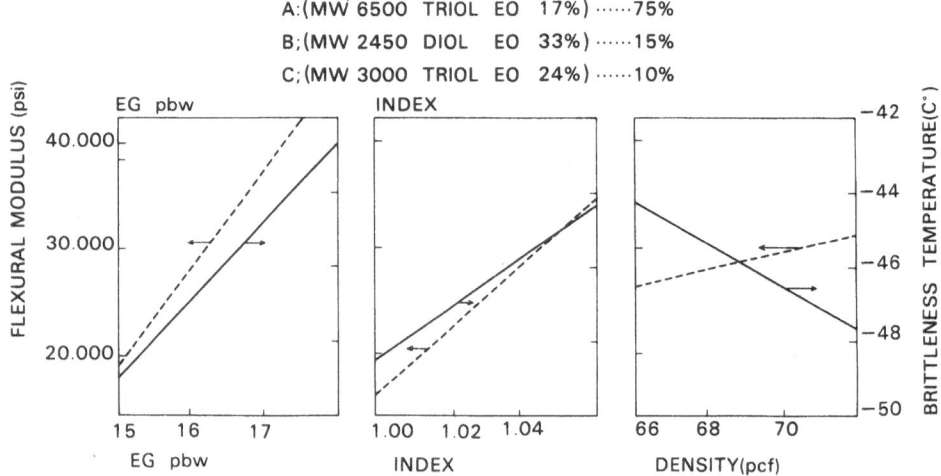

Figure 3.  Correlation of physical properties by regression model.

9000 psi and the brittleness temperature rise by about 3°C for one part of ethylene glycol per total weight of polyol.  Figures 4 and 5 show the relationship between the flexural modulus and the brittleness temperature on the polyol composition.  Other physical properties were obtained by the same techniques, but they are not shown here. The system which fits the objective physical properties can be obtained with these figures.

METHOD FOR THE SCREENING OF POLYOL COMPOSITION

The method for the screening of the polyol composition which fits the objective physical properties is as follows:

1.  Selection of an isocyanate.

2.  Selection of two or three polyols taking into account the objective physical properties.

3.  Screening five or ten formulations for various chain extender levels and polyols levels.

4.  Molding (the density and index should be changes).

5.  Measuring the physical properties.

6.  Regression analysis (BMD-02D).

7.  Investigation over the systems by use of regression equation.

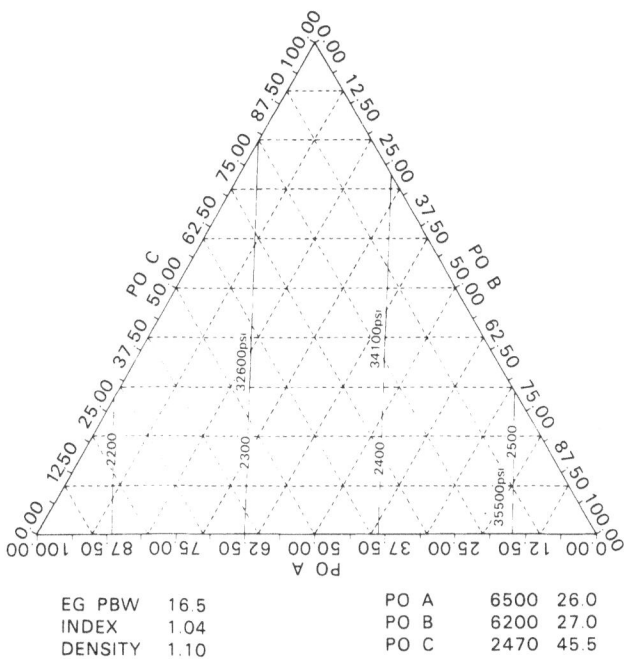

| EG PBW | 16.5 | PO A | 6500 | 26.0 |
| INDEX | 1.04 | PO B | 6200 | 27.0 |
| DENSITY | 1.10 | PO C | 2470 | 45.5 |

Figure 4.   Dependence of flexural modulus on the composition of
            polyols.  (16.5 pbw ethylene glycol.)


     8.   Determination of the optimum system.

The qualitative effect of the molecular weight and the primary OH
value of polyol upon the physical properties can be determined by
using described regression analysis.


EVALUATION METHOD FOR RIM MOLDABILITY

     Not only physical properties but also moldability should be
taken into account for the production of an adequate RIM system.
The moldability of RIM is strongly affected by the molecular weight,
the number of functional groups and the ratio of primary OH in the
end group of polyol, type of isocyanate, catalysts, blowing agents,
isocyanate index, etc.  The best system for physical properties is
not always the best one for moldability.  For example, the higher
the molecular weight of the polyol, the poorer is the green strength.
The system with both good flowability and good reactivity is required for high productivity.  For that reason the adequate estimation method for moldability has been pursued.  Until this time we
have had to depend on these clearly uncertain methods and the sub-
jectivity of the person doing the experiments.  An absolute estima-

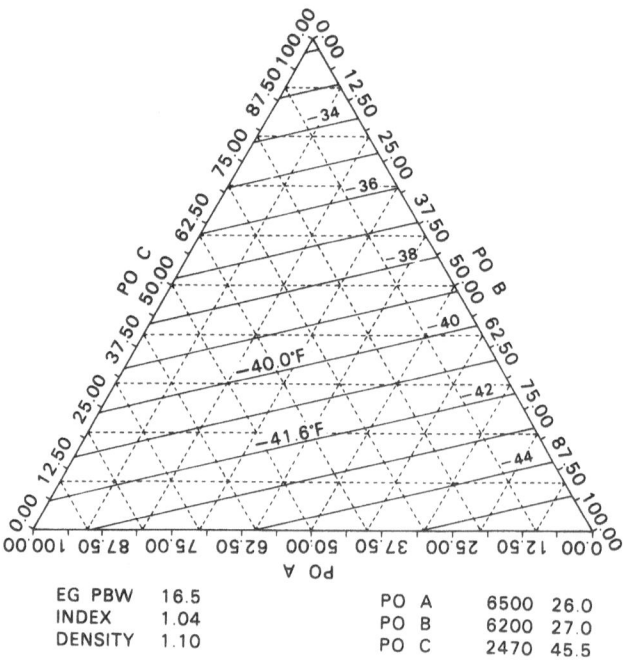

| EG PBW | 16.5 | | PO A | 6500 | 26.0 |
| INDEX | 1.04 | | PO B | 6200 | 27.0 |
| DENSITY | 1.10 | | PO C | 2470 | 45.5 |

Figure 5.   Dependence of the brittleness temperature on the
            composition of polyols.  (16.5 pbw ethylene glycol.)

tion method for moldability has long been desired.  It was clear from
Domine's studies that there was distribution of conversion to the
flow direction.  Therefore we estimated that there still remained
the characteristics of the system in the density distribution.


EQUIPMENT AND EXPERIMENTAL CONDITIONS

     The RIM machine Admiral 2000 and the step-mold for the investi-
gation of the moldability were used in this study.  The step-mold had
three parts of thickness 2.0 mm, 2.5 mm and 3.0 mm.  The mold was
55 cm in width and 100 cm in length.  The volume of the mold was
1390 cm$^3$.  The mold temperature was precisely kept at 70±1 °C.
Systems based on grafted polyols, 1,4-butanediol and various prepolymers were used.  DBTL was used as catalyst.  Three different
batches (standard, ± 100 g) of the material were used for one system
and their actual densities were measured.


COMPARISON OF THE MOLDABILITY BY DENSITY DISTRIBUTION

     The product was divided into twenty-eight pieces as shown in

Figure 6.  The distributions were measured for each piece and the density contour lines were recorded by an X-Y plotter.  Figure 6 shows the density contours when the amount of the materials in the mold was varied.  The density contour lines drawn at every 0.01 g/cm$^3$ were very different depending on the amount of the materials used. The density distribution generally decreased with the increase of the amount of loaded material.  Therefore, the constant loading of the materials is needed for the density distribution study.

Figure 7 shows the density contours when various prepolymers were used with one polyol system.  System A shows the uniform distribution and seems to have the best flowability, while system D seems to have the worst flowability.  We considered this density distribution as a projection on the plane, which shows the characteristics of the reaction.  Material loads of three different weights for one system were used to clear up the condition of the reaction.  We considered the next four scales to recognize the

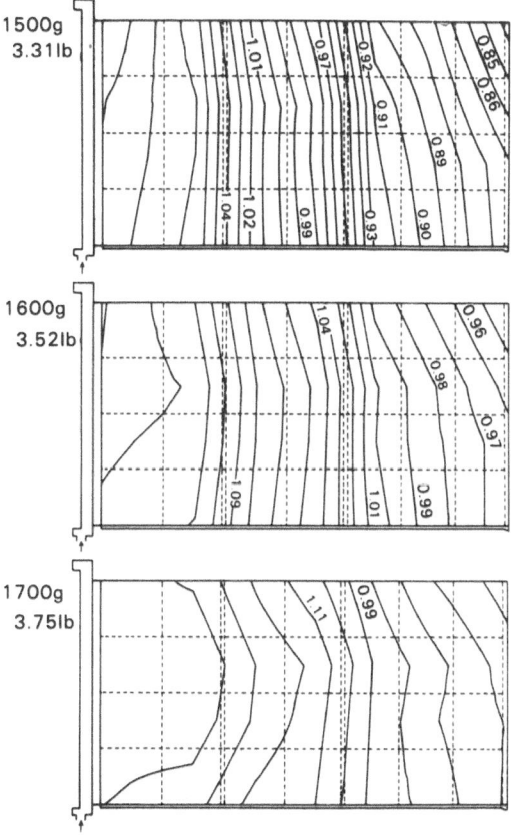

Figure 6.  Dependence of the density distribution on the mold loading.

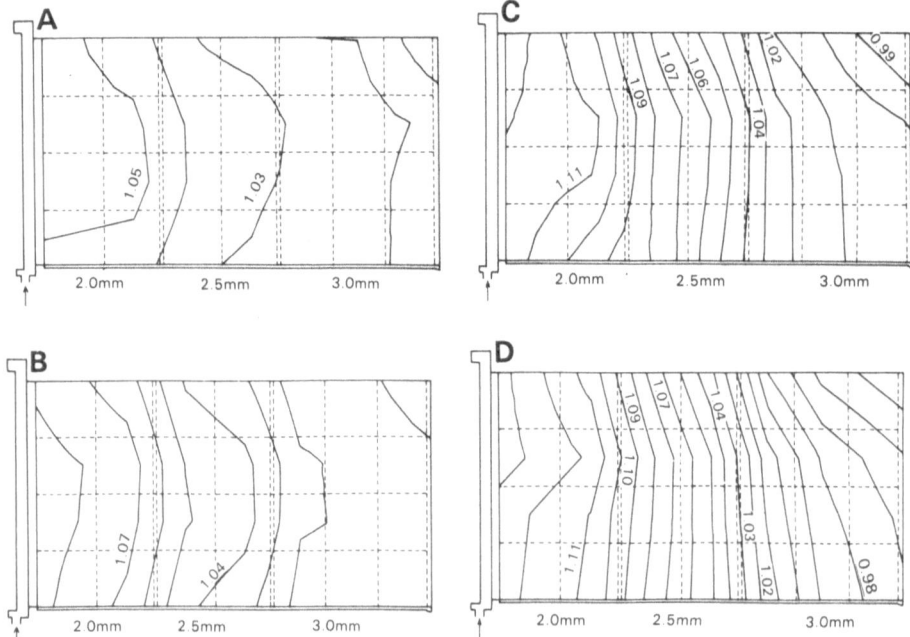

Figure 7.  Density distribution contours for various RIM systems.


patterns on the plane.

## Variance of Density Distribution

The density has distributions for volume or area of the product. The degree of the distribution was obtained by calculating the variance of density. This value was calculated by using Equation 3.

$$VAR = \frac{1}{N} \Sigma a_i \, D^2_i - \overline{D}^2_i$$
$$N = \Sigma a_i \tag{3}$$

## Center of Gravity

The center of gravity was easily calculated by using Equation 4. The dimensionless values were obtained by Equation 5 when the origin was the product end near the injection side.

$$X = \frac{\Sigma W_i x_i}{\Sigma W_i} \qquad Y = \frac{\Sigma W_i y_i}{\Sigma W_i} \tag{4}$$

$$COG = \frac{(X^2 + Y^2)^{0.5}}{(X_o{}^2 + Y_o{}^2)^{0.5}} \tag{5}$$

## Average Density Point

The curve which corresponds to the value of the overall average density varied according to the pattern of the flow. The point where this curve intersects the center line of the product was calculated by Equation 6.

$$ADP = \frac{(X'^2 + Y'^2)^{0.5}}{(X_o{}^2 + Y_o{}^2)^{0.5}} \tag{6}$$

## Distribution of Increment of Weight

Equation 7 was used to calculate the index of the flow. The ratio of the increment near the gate and the increment of the flow end was calculated by using Equation 8.

$$Z_{ij} = \frac{wij - wij'}{\Sigma wij - \Sigma wij'} \times 100 \tag{7}$$

$$wI = \frac{\Sigma Z_{7j}}{\Sigma Z_{1j}} \quad \text{(weight increasing far from the gate)} \atop \text{(weight increasing near the gate)} \tag{8}$$

The results for eight systems are shown in Table V. Eleven variables for each system can be obtained by varying the load amounts of the materials. These data could be expected to express the previously described moldability. But as these data involve two factors of moldability, it is difficult to clearly distinguish the characteristics of moldability. We considered the Principal Component Analysis to solve this problem.

## Results of Principal Component Analysis

The principal component analysis helps to summarize a number of correlative data into a few non-related characteristics. We applied this method to twenty-four systems, and could abstract 89% of the information by two principal components from the original data. This means that the original eleven variables were summarized into two independent variables. The results are summarized in Table VI. The physical meaning of each component is usually understood by considering the eigenvectors between principal components and original variables.

Table V.  Density Distribution Data

| SYSTEM NO. | VARIANCE OF DENSITY | | | CENTER OF GRAVITY | | | AVERAGE DENSITY | | | INCREMENT | |
|---|---|---|---|---|---|---|---|---|---|---|---|
| | 3.31 lb | 3.52 lb | 3.75 lb | 3.31 lb | 3.52 lb | 3.75 lb | 3.31 lb | 3.52 lb | 3.52 lb | 3.3-3.5 lb | 3.5-3.7 lb |
| 1 | 665 | 299 | 102 | 0.964 | 0.978 | 0.987 | 1.016 | 1.041 | 1.083 | 4.72 | 3.75 |
| 2 | 130 | 50 | 21 | 0.983 | 0.989 | 0.991 | 1.020 | 1.030 | 1.079 | 2.14 | 2.82 |
| 3 | 154 | 55 | 26 | 0.982 | 0.987 | 0.989 | 0.998 | 1.022 | 1.018 | 2.38 | 2.25 |
| 4 | 38 | 12 | 18 | 0.993 | 0.993 | 0.987 | 1.045 | 1.037 | 1.068 | 1.81 | 1.41 |
| 5 | 567 | 188 | 65 | 0.967 | 0.982 | 0.987 | 0.976 | 1.013 | 1.040 | 9.18 | 3.92 |
| 6 | 1020 | 343 | 86 | 0.956 | 0.974 | 0.987 | 1.029 | 1.037 | 1.091 | 18.46 | 5.06 |
| 7 | 1279 | 521 | 79 | 0.948 | 0.972 | 0.987 | 1.020 | 1.052 | 1.107 | 11.97 | 6.02 |
| 8 | 1305 | 517 | 160 | 0.948 | 0.971 | 0.982 | 1.085 | 1.087 | 1.104 | 11.29 | 6.45 |

Table VI.  Results of Principal Component Analysis

| | Proportion of total variance | Cumulative proportion |
|---|---|---|
| $Z_1$ : Component 1 | 75 | 75 |
| $Z_2$ : Component 2 | 14 | 89 |
| $Z_3$ : Component 3 | 6 | 95 |

The first principal component is estimated to have high correlation with flowability.  It is clear from the eigenvectors that this component has high correlation with the variance of density and the center of gravity.  The second principal component has high correlation with the average density point and the increment of weight at the low loading of the materials.  This second principal component is estimated to have something to do with the reactivity.

Both sink marks relating to flowability and the loose skins related to reactivity were observed to make the meaning of the principal components clearer.  Figure 8 shows the relationship between the first principal component Z1 and the percentage of the area of the sink marks in the first 2.0 mm plane.  The area of sink marks decreases with the value of Z1; thus it is estimated that Z1 means the flowability.  The second principal component Z2 is strongly related with the area of the loose skins as shown in Figure 9.  Considering the eigenvectors (Table VII), this principal component is estimated to mean the reactivity.  Figure 10 shows the characteristics of the systems by plotting the first principal component as the ordinate and the second principal component as the abscissa.  The system in the direction of quadrant 'I' has better flowability and better reactivity.  These systems will fit the large and the thin type of bumper.  The Quadrant 'IV' is the zone where the reactivity is best but the flowability is poor.  The increase of the

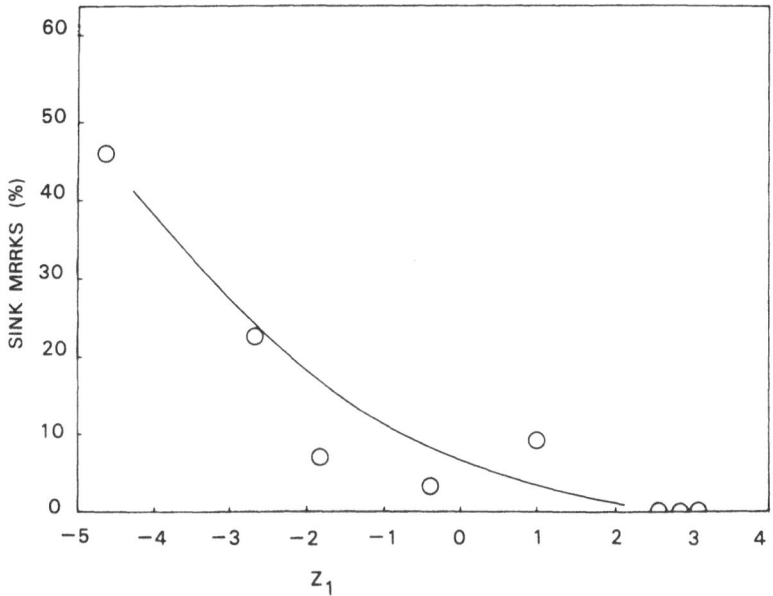

Figure 8. Dependence of the area of sink marks measured in 60 segments of the first 2.0 mm of product on the factor loadings of the first principal component.

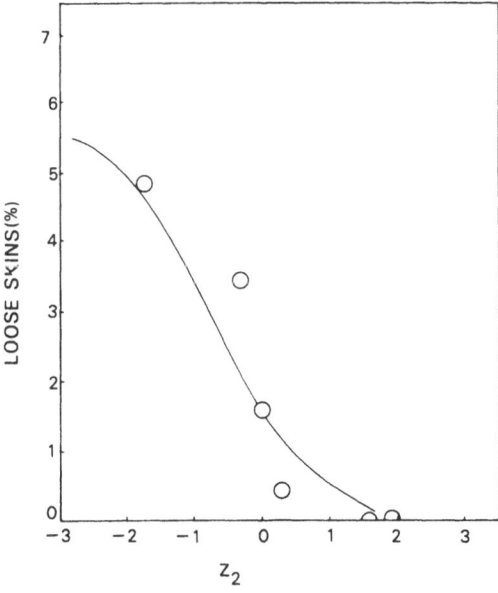

Figure 9. Dependence of the area of loose skins measured in 60 segments of the first 2 mm of product on the factor loadings
of the second principal component.

Table VII.  Eigenvectors of Principal Components Z1 and Z2

$$Z_1 = \sum b_{1i} \cdot wi$$
$$Z_2 = \sum b_{2i} \cdot wi \quad \cdots\cdots\cdots\cdots ( 9 )$$

| $w_i$ | $Z_1$ $b_{1i}$ | $Z_2$ $b_{2i}$ |
|---|---|---|
| VAR 1 | −0.3409 | −0.1528 |
| VAR 2 | −0.3391 | −0.1080 |
| VAR 3 | −0.3219 | 0.0548 |
| WI 1 | −0.2692 | −0.3325 |
| WI 2 | −0.3326 | −0.1815 |
| COG 1 | 0.3302 | 0.2439 |
| COG 2 | 0.3322 | 0.2213 |
| COG 3 | 0.2676 | −0.2547 |
| ADP 1 | −0.1958 | 0.6323 |
| ADP 2 | −0.2813 | 0.4480 |
| ADP 3 | −0.2715 | 0.2110 |

catalyst DBTDL by 0.02 pbw for the same system caused the movement of point A to positions B and C.  The catalyst DBTDL improves the reactivity but deteriorates the flowability within the limit of this catalyst level.


DISCUSSION OF RIM MOLDABILITY USING MODEL DATA

The twelve model data are shown in Figure 11 to clarify the relationship between the system and the moldability.  Each figure expresses dependence of the density distribution on the flow length. The three lines show the patterns when the amount of loaded material is varied.  The density distribution increasingly declines from pattern A to pattern H.  The variance of density is the lowest at pattern A.  The three density distribution lines are parallel in patterns A, D and H.  The pattern H has steeper density distribution than patterns D and A.

These model data plotted on the coordinates of Z1 and Z2 are shown in Figure 12.  The movement of the patterns H, D and A in the figure makes it clear that the flowability becomes better when the reactivity is nearly constant.  The movement of the patterns I, D and J is very effective in increasing the reactivity.

The changes of the patterns I, D and J keep the average density point away from the gate.  It becomes clear that the more reactive systems have more gentle density distribution near the gate within the limits of our data.  This is similar to the trend seen in Broyer's paper, wherein it is shown that the more reactive systems easily reduce the viscosity in the earlier part of the reaction period.

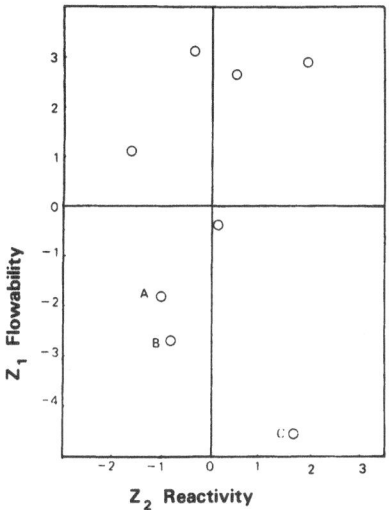

Figure 10.  Evaluation of moldability by principal component analysis.

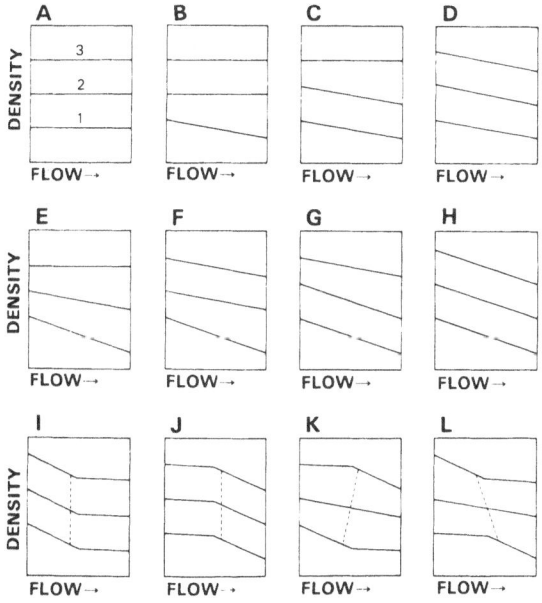

Figure 11.  Model patterns.

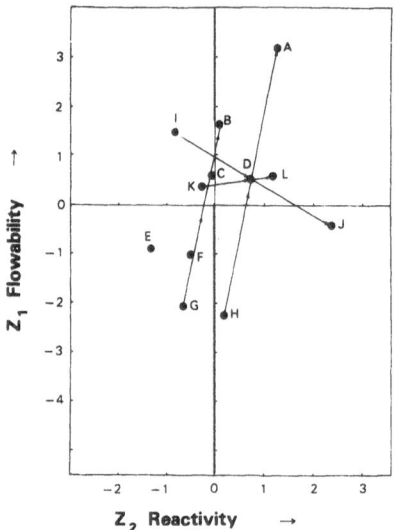

Figure 12.   Plot of the flowability data Z1 on reactivity by Z2
             for the model system data.


     The most adequate system for RIM is similar to pattern A, which
has low and even density distribution when the average density point
is kept away from the gate.   This system will be expected to give
high productivity for large and thin types of bumpers.


CONCLUSIONS

     The analysis of the density distribution utilyzing the principal
component analysis helped us understand the moldability quantita-
tively.   The procedure for the estimation of moldability is as
follows:

     1.   Molding of the samples for various systems.

     2.   Measurement of densities.

     3.   Calculation of estimation variables.

     4.   Principal component analysis of estimation variables.

     5.   Plotting on the coordinate of principal components.

     6.   Evaluation of each system.

     7.   Determination of the optimum system.

It must be understood, however, that these statistical methods are
accurate within the limits of the data but beyond the data they
might become questionable.  In order to reach a more generalized
theory, a great number of data must be accumulated.  The RIM moldability will become more clear by combining the moldability studies
with these simulation methods.  These methods can be successfully
used for the screening of chemical systems in RIM applications.


Nomenclature

$POH$ = primary OH ratio in the end group of polyol

$MW$ = molecular weight of polyol

$OH$ = OH value of polyol

$Wi$ = weight fraction of polyol-i

$Di$ = density of the piece-i in product

$ai$ = area of product occupied by density Di

$(X,Y)$ = coordinates of center of gravity in product

$(X',Y')$ = coordinates of average density point in product

$COG$ = dimensionless value of center of gravity

$ADP$ = dimensionless value of average density point

$Zij$ = increment of weight at the piece-in of product

$Wij$ = weight of the piece in i-th row, j-th column

$W'ij$ = less loaded product's one

$WI$ = ratio of the increment of weight

$Zi$ = i-th principal component

$wi$ = standardized original variables

$bi$ = eigenvectors of principal component

$(X_0,Y_0)$ = coordinates of center of gravity at constant density

References

1.  P. S. Carleton, J. H. Ewen, T. M. Shah, E. J. Thompson, H. E.
    Reymore, Jr., and A. A. R. Sayigh, SAE paper (1974).
2.  J. D. Domine and C. G. Gogos, SPE Tech. Papers, $\underline{22}$, 274 (1976).
3.  E. Broyer, C. W. Macosko, F. E. Critchfield and L. F. Lawler,
    Polymer Eng. Sci., 18, 5, 382 (1978).
4.  P. S. Carleton, E. J. Thompson, T. M. Shah, H. E. Reymore, Jr.,
    and A. A. R. Sayigh, J. Cellular Plastics, Jan/Feb, 36 (1977).
5.  E. K. Moss, J. Cellular Plastics, Sep/Oct, 282 (1969).

# INDEX